Ermakova A.V.

ADDITIONAL FINITE ELEMENT METHOD FOR ANALYSIS OF REINFORCED CONCRETE STRUCTURES AT LIMIT STATES

ASV Construction
Stockholm, Sweden
2016

Universal Decimal Classification: 624.15.04(075.8)

Reviewers:

Professor Golovin N.G., the Head of Department of Reinforced Concrete and Stone Structures of Moscow State Building University, Doctor of Sciences.
Dr. Krilov S.B., a Chief Research Fellow, Doctor of Technical Sciences.

Ermakova A.V.
Additional Finite Element Method for Analysis of Reinforced Concrete Structures at Limit States / ASV Construction, 2016. – 116 p.

ISBN 978-91-982223-2-6

The work presents the theoretical basis of Additional Finite Element Method (AFEM), which is a variant of the Finite Element Method (FEM) for analysis of reinforced concrete structures at limit state. AFEM adds to the traditional sequence of problem by FEM the units of the two well-known methods of the structural design: method of additional loads and limit state method. The problem is solved by introduction of ideal failure models and additional design diagrams formed from additional finite elements, where each AFE describes the limit state reached by the main element. The main relations defining the properties of AFEs as well as the examples of the use of Additional Finite Element Method for analysis of reinforced concrete structures at limit state are given in the work too.

Translated from the Russian by Ermakova O. V.:

ISBN 978-91-982223-2-6

CONTENTS

INTRODUCTION

Nowadays finite element method (FEM) is the most prevalent and efficient method of analysis of building structures in our country and abroad [1,3–5,23,43–45,48,50,52–55,63,64,68,76,84–86,88,93,115,120,122,123].

In.1. The solution of linear problem by finite element method

The main point of analysis of any structure is that it is considered as an assembly of definite number of parts with finite size and simple form – finite elements. It allows to reduce an analysis to the solution of the system of algebraic equations.

In.1.1.The main stages of analysis of reinforced concrete structures by finite element method

The next stages are necessary for solution of any problem by finite element method:

1. Composition of a design diagram of the structure, i.e. a representation of the structure in the form of assembly of separate finite elements (FEs). This operation includes an assignment of size, shape of finite elements and the way of their joint. Usually the finite elements are jointed in the definite points – nodes.

2. Calculation of a stiffness matrix of the separate finite element. This matrix determines the connection between node forces (stresses) and corresponding node displacements:

$$R = K_e V, \qquad \text{(In.1)}$$

where R = vector of node reactions of the finite element;
V = vector of node displacements of the finite element;
K_e = stiffness matrix of the given finite element.

3. Formation of a decision system of linear algebraic equations from the equations of equilibrium at each node of the considered structure with node displacements as the unknowns:

$$K V = P, \qquad \text{(In.2)}$$

where V = matrix–column of unknown node displacements;
P = matrix–column of external load;
K = stiffness matrix of the considered structure. This matrix is formed from coefficients of stiffness matrices of the separate finite elements.

4. Calculation of node displacements by means of solution of the system of linear algebraic equations:

$$V = K^{-1} P, \qquad \text{(In.3)}$$

where K^{-1} = inverse stiffness matrix of the considered structure.

Most commonly Gauss elimination of unknowns is used for solution of the system of linear algebraic equations. This means that the step–by–step elimination of unknowns in the system of equations is carried out at fist beginning with the second equation up to the last one (direct execution), and then the step-by-step determination of unknowns is carried out beginning with the last equation up to the first one (reverse execution).

5. Calculation of strains, stresses and node reactions according to the obtained displacements.

6. Analysis of stress-strain state of the structure.

If the calculation is carrying out within the limits of solution of linear problem, i.e. in an elastic behavior of structural material for one type of loading, then after the fulfillment of the enumerated above six stages it may be considered a completed one. The enlarged design diagram of linear analysis by finite element method is given in Fig. In.1. The calculation should be repeated several times to determine the structural stress-strain state at some steps or schemes of load.

Design diagram of linear analysis of structure by finite element method

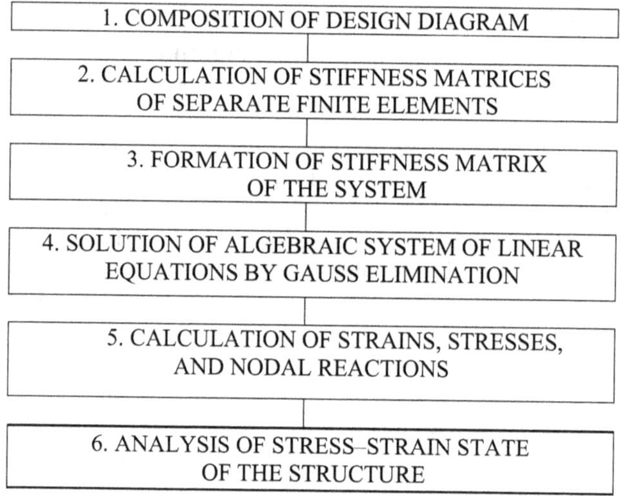

1. COMPOSITION OF DESIGN DIAGRAM
2. CALCULATION OF STIFFNESS MATRICES OF SEPARATE FINITE ELEMENTS
3. FORMATION OF STIFFNESS MATRIX OF THE SYSTEM
4. SOLUTION OF ALGEBRAIC SYSTEM OF LINEAR EQUATIONS BY GAUSS ELIMINATION
5. CALCULATION OF STRAINS, STRESSES, AND NODAL REACTIONS
6. ANALYSIS OF STRESS–STRAIN STATE OF THE STRUCTURE

Fig. In.1.

However it should be noted that in most cases the linear analysis does not give sufficiently reliable information about the behavior of structure. The obstacle is a physical and geometrical nonlinearity of structural behavior under load. In this case a number of sequential steps usually are fulfilled to obtain a real picture of the structural stress-strain state.

In.1.2. The main types of finite elements

Analysis of any structure begins with a preparation of its design diagram, i.e. with the presentation of the structure in the form of assembly of finite elements. This operation includes some interacting steps: definition of a type of the solving problem, choice of a type of the simulating finite element and partition of the structure by finite elements.

Nowadays most often next systems are analyzed by finite element method: plane and spatial frames, bending slabs, deep-beams, shells, plane and spatial systems composed of various working simultaneously structural members.

Usually the initial point of a definition of the problem type is the number of degrees of freedom, i.e. the number of possible displacements of each node of the considered structure, which in turn is determined by the stress-strain state of the structure from the point of view of structural mechanics. The simulating finite element and its size are chosen after determination of behavior of the structure as a whole starting from the necessary number of displacements of each its node.

Thus most often next types of finite elements are used for simulation of structural members: "bar of general attitude", "deep-beam", "bending slab", "plane element of the shell of general attitude" [52, 64,65,84,85]. In choice of simulating finite elements it is necessary to watch strictly that the nodes of the chosen finite elements and the structure as a whole ought to have the same degrees of freedom: otherwise the problem has no decision.

In addition special finite element named "connecting" (CFE) without size but with definite stiffness is used for taking into account of node flexibility of the structure and for simulation of behavior of soil base. It allows to take into account the singularities of structural behavior analyzed by finite element method more substantive.

In.1.3. The triangular deep-beam finite element with linear properties

The triangular deep-beam finite element with linear properties (Fig. In.2) is destined for analysis of thin plane stresses systems situated in plane *XOY* [45, 51,81,86,88,115]. Each of the three nodes *i*, *j*, *k* of the finite element has two degrees of freedom, i.e. two possible linear displacements *w* and *u* along axes *X* and *Y* respectively. As a result of analysis the values of node displacements of finite element are calculated by means of which next stresses in the center of gravity of finite element are determined (Fig. In.3): σ_x – direct stress acting along axis *X*; σ_y – direct stress acting along axis *Y*; τ_{xy} – tangential stress.

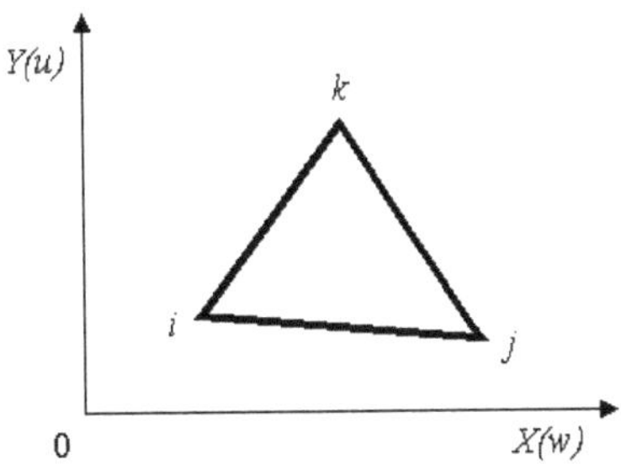

Fig. In.2. The triangular deep-beam finite element with linear properties

The relationship between node reactions R and node displacements V is determined by the formula (In.1).

The stiffness matrix K_e of this finite element in system of axes XOY is calculated according to familiar formula:

$$K_e = tS(A^{-1})^T B^T DBA^{-1}, \qquad \text{(In..4)}$$

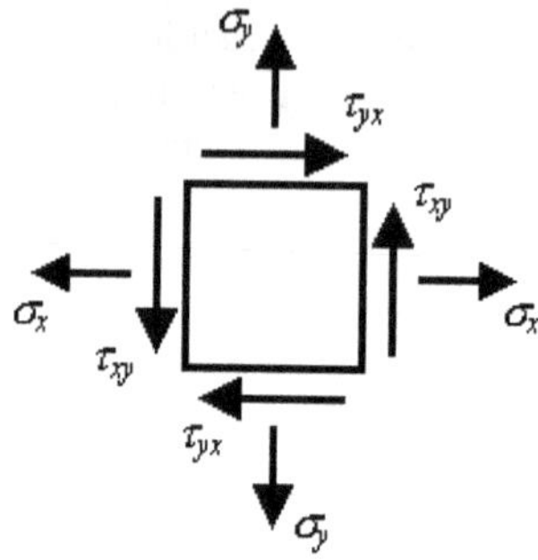

Fig. In.3. Stresses in the centre of gravity of triangular deep-beam finite element

The expanded form of the formula (In.1) is:

$$R = tS(A^{-1})^T B^T DBA^{-1} V, \qquad \text{(In.5)}$$

where t = thickness of finite element; S = area of finite element; A = matrix of node coordinates; A^{-1} = inverse of the matrix A; $(A^{-1})^T$ = transposed matrix A^{-1}; B = matrix connecting node displacements and stresses of finite element; B^T = transposed matrix B; D = matrix of elasticity.

The relationship between node displacements V of finite element and its strains ε has the form:

$$\varepsilon = BA^{-1} V. \qquad \text{(In.6)}$$

The relationship between stresses σ and strains ε is:

$$\sigma = D\varepsilon. \qquad \text{(In.7)}$$

The relationship between node reactions R and stresses σ is:

$$R = tSC\sigma, \qquad \text{(In.8)}$$

where $C = (A^{-1})^T B^T$.

The relationship between node reactions R and strains ε is:

$$R = tSG\varepsilon, \qquad \text{(In.9)}$$

where $G = (A^{-1})^T B^T D = CD$.

Formulae (In.1), (In.4)–(In.9) describe the properties of classical triangular linear deep-beam finite element completely.

In.1.4. The connecting finite element

Connecting finite element (CFE) simulating bond between adjacent nodes is destined for taking into account of pliability of joints [52, 63, 65,120]. In analysis of deep-beams or plane frames each node s or b of the finite element has two degrees of freedom, i.e. two possible displacements w and u along axes X and Y (Fig. In.4). This element has two elastic bonds of finite stiffness.

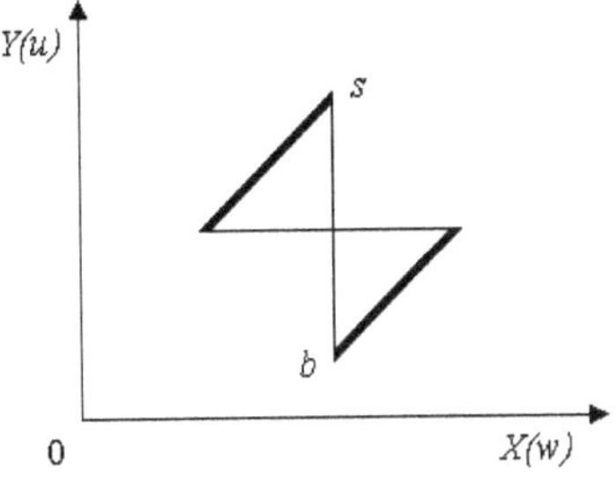

Fig. In.4. The connecting finite element (CFE) simulating elastic bond between two nodes

The relationship between its node reactions R and node displacements V is determined by formula (In.1). At that time for this finite element it is better to use the link between stresses in bonds T and mutual node shifts g, which is in the form of the next formula:

$$T = K_e\, g\,, \tag{In.10}$$

where K_e = bond rigidity.

Mutual shifts of node s and b are calculated by the formula:

$$g = V_s - V_b, \tag{In.11}$$

where V_s, V_b = shifts of nodes s and b respectively in the connecting finite element.

Formulae (In.1), (In.10) and (In.11) characterize the properties of the given connecting finite element completely.

In.1.5. The finite element "Bar of general attitude"

The finite element "Bar of general attitude" is destined for analysis of arbitrary spatial bar systems [52,64,84,85]. Each node has six degrees of freedom, i.e. three possible linear displacements along coordinate axes and tree angular displacements around the same axes (see Fig. In.5). The next system of stresses characterizes this finite element: N – longitudinal force; M_y – bending moment relative to one axis of inertia; M_z – bending moment relative another axis of inertia; M_k – twisting moment relative longitudinal axis of bar; Q_y – shear force acting along one axis of inertia; Q_z – shear force acting along another axis of inertia.

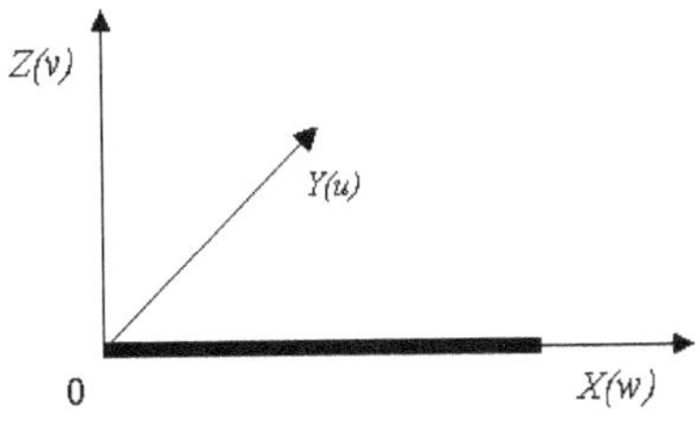

Fig. In.5. The finite element "Bar of general attitude"

In.2. Solution of nonlinear problem by finite element method

In most cases linear analysis of structures by finite element method does not give enough real information about stress-strain state of these structures due to their strongly expressed nonlinear properties. Therefore the solution of nonlinear problem by finite element method remains an actual subject.

In.2.1. Two ways for the allowance of the nonlinear properties of materials

As it was mentioned above, finite element method in the classical linear statement reduces the solution of any problem to the solution of a system of linear equations (see formula In.2). For example in analysis of reinforced concrete structures such linear approach does not give real results as well as reinforced concrete has a number of specific nonlinear properties, in particular strongly expressed plastic properties: cracking, bond between concrete and reinforcement, prestressing and so on. In view of it the analysis is carried out by step-by-step method of gradual increase of load often with several iterations at each step. It is necessary to take into account precisely the typical nonlinear properties of the structure and the degree of their manifestation in the considered stage of loading.

Thus the presence of strongly expressed nonlinear properties of reinforced concrete has required to develop the special ways for the allowance of them in analysis of reinforced concrete structures.

Two kinds of such ways may be distinguished. The first kind is connected with the change of coefficients of stiffness matrix K at each step or iteration and the second one is connected with the change of vector of external load P.

Appearance of these kinds reflects the two opportunities of influence on the results of solving of the system of linear algebraic equations (1.1): by change of the left part of the system by variation of coefficients of unknowns, in this case of elements of the general stiffness matrix of the structure K [3,48,84,85], and by change of the right part, i.e. the column of constant terms, in this case of vector of external load P [86]. Moreover both in that case and in another one the fulfilled changes ought to be logically connected with the correction of formulae for determination of stresses in finite elements.

As it was mentioned above the solving of the system of linear algebraic equations is realized by Gauss method. According to this method the unknowns are determined at first by direct execution, then by reverse execution. Operation of direct execution includes sequential elimination of unknowns, i.e. operations of calculation of inverse stiffness matrix of the structures K^{-1} and connected with it operations of transformation of column of constant terms P. Obtaining of inverse stiffness matrix is the most laborious operation from the operations of direct execution: it takes roughly three quarters of the time necessary for solving of the system as a whole. Operation of reverse execution includes sequential calculations of unknown node displacements V, beginning with the last one. These operations are less laborious than the operations of direct execution.

In the large if we consider the structural analysis by finite element method, it should be pointed out that the solving of the system of linear equations (In.2) is the most laborious and prolonged stage of problem solving which determines the whole of process.

In.2.2. The first way for the allowance of the nonlinear properties

The first way for the allowance of the nonlinear properties of material of the structure connecting with the change of coefficients of general stiffness matrix *K* at each step of loading or iteration presupposed that every time the matrix is composed anew and the system of linear equations (In.2) is solved again [10,24,89,98,109,116].

At present this way is realized in most programs using finite element method [3,48,84,85]. The cause is that their use does not require reorganization of the main algorithm of the problem solving. Most commonly it is sufficient to insert necessary change in carrying out of operations preceding the solution of the system of linear equations (see Fig. In.6). Thus, to take into account plastic properties it is possible to get along by assignment of nonlinear law of material deformation and sometimes it is enough to change its modulus of elasticity *E*.

Design diagram of structural analysis by finite element method for the allowance of the nonlinear properties according to the first way

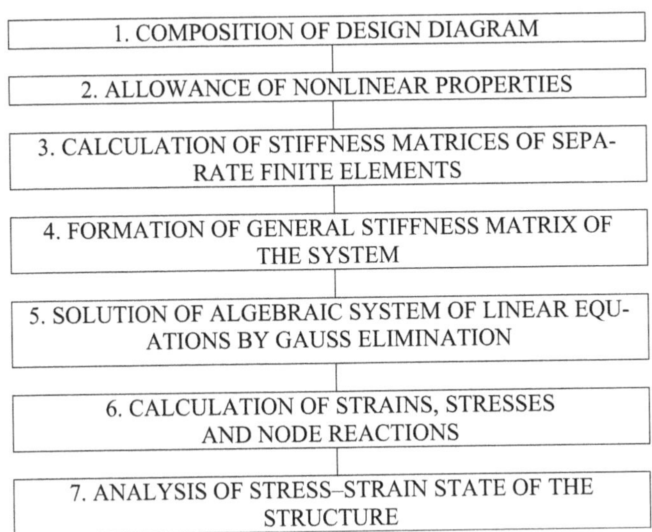

1. COMPOSITION OF DESIGN DIAGRAM
2. ALLOWANCE OF NONLINEAR PROPERTIES
3. CALCULATION OF STIFFNESS MATRICES OF SEPARATE FINITE ELEMENTS
4. FORMATION OF GENERAL STIFFNESS MATRIX OF THE SYSTEM
5. SOLUTION OF ALGEBRAIC SYSTEM OF LINEAR EQUATIONS BY GAUSS ELIMINATION
6. CALCULATION OF STRAINS, STRESSES AND NODE REACTIONS
7. ANALYSIS OF STRESS–STRAIN STATE OF THE STRUCTURE

Fig. In.6.

Nowadays the change of the general stiffness matrix of the structure usually is carrying out by next rather simple ways: a) variation of input data, for example, stiffness characteristics of finite elements of the structure design diagram; b) variation of stiffness matrices of finite elements of the given system , for example, by introducing of zero finite elements or specially corrected ones directly in these matrices (usually it is carried out at the stage of calculation of stiffness matrices of the separate finite elements before formation of general stiffness matrix of the structure K); c) variation of design diagram of the structure by transformation, for example, by node dividing or introducing of new finite elements with special properties.

Usually even in analysis of a simple structure it is impossible to be restricted to the use of any single way for the allowance of the nonlinear properties of the structure, since these properties are various by their nature and character of influence on behavior of the structure.

In.2.3. The second way for the allowance of the nonlinear properties

The second way for the allowance of the nonlinear properties of material of the structure connecting with change of vector P at each step or iteration is named as the method of additional loads [28, 29, 50, 86]. This means that if we use this method the system of linear algebraic equations (In.2) has the next form:

$$K\,V = P + F, \tag{In.12}$$

where F = vector of additional load taking into account nonlinear properties.

In.2.3.1. Method of additional loads

The second way is based on method of additional loads (elastic decision) suggested by Ilyushin A.A. [47] for solving of the deformational problems of plastic theory. According to this method in place of elastic-plastic body an ideal elastic one with the same strains but with additional loads is taken. Therefore this method is called the method of additional loads. It is based on separation from the stiffness matrix of its linear component [81]:

$$K_{nonl} = K + \Delta K_{nonl}, \tag{In.13}$$

where K_{nonl} = stiffness matrix with nonlinear properties;

K = the linear part of stiffness matrix K_{nonl};

ΔK_{nonl} = the nonlinear part of stiffness matrix K_{nonl}.

If we substitute the expression (1.13) in the formula (1.2), remove the parentheses, transfer the second term in right hand side, we may get the next expression:

$$K\,V = P - \Delta K_{nonl}\,V. \tag{In.14}$$

After comparison of expression (In.12) with (In.14), we may calculate the vector of additional loads by the next formula:

$$F = -\Delta K_{nonl}\,V\,. \tag{In.15}$$

This formula allows to determine the value of additional load in carrying out of the process of iteration. In this case the value of the vector of node displacement V is obtained from the previous iteration. The design diagram of elastic decision method for the system with one degree of freedom is shown in Fig. In. 7a.

This method allows to solve the algebraic system of linear equations (In.12) with constant coefficients in the right hand side, i.e. by obtaining of inverse matrix K^{-1} only once. The main advantage is contained in its simplicity.

As is obvious from Fig. In. 7 the necessary number of carrying out iterations depends on the accuracy of the first linear approximation. Therefore the representation of this approximation determines the whole of problem solving.

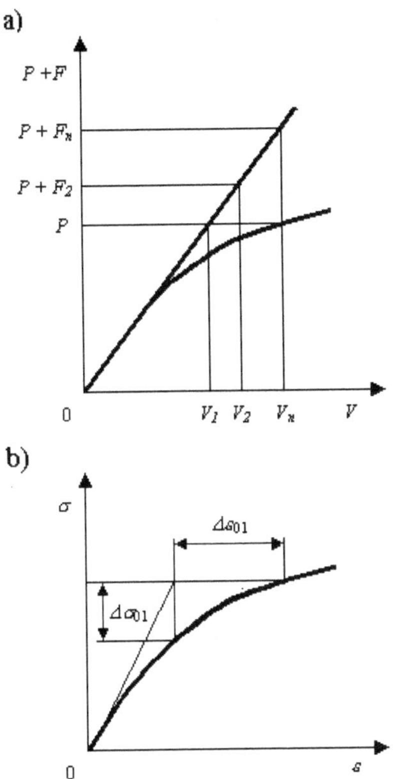

Fig. In.7. Methods of elastic solutions, initial deformations and initial stresses: a) scheme of the method of elastic solutions[80]; b) scheme of the methods of initial deformations and initial stresses[43]

It should be pointed out that this method has two considerable shortages: a) the process of iteration badly converges with great difference between linear and nonlinear solving, i.e. with nonlinear deformations; b) the process of iteration gives the oscillating solution when the curve $(P + F)$ has some definite form.

The method of elastic solutions is closely connected with two other methods [9,67,119] derived from it: method of initial (additional) deformation and method of initial (additional) stresses (Fig.In.7b) [43]. Using the method of initial deformation at each step of achieved stresses we determine the initial deformation $\Delta\varepsilon_{01}$, which is used for formation of the vector of additional load F [8,67]. The method of additional stresses is analogous to the previous one but the initial stresses $\Delta\sigma_{01}$ are used for formation of the vector of additional load F.

The final choice of the way of additional stresses formation is determined by the conditions of each particular problem.

In.2.3.2. Singularities of the combination of finite element method and method of additional loads

At present the second way is used in a limited number of programs which are realized the analysis of the structures by finite element method. The cause is that the use of these methods requires of essential modification of the algorithm of the problem solution at the main stage – the solving of algebraic system of linear equations, i.e. the construction of the program ought to be according to the design diagram given at Fig. In.8. Besides there is a serious problem of obtaining of theoretical relationships which allow to form the vectors of additional loads and connecting with them the vectors of correcting stresses in finite elements for realization of the method of additional loads in the software packages destined for nonlinear analysis of the structure by finite element method.

However namely the use of this way allows to eliminate the most labor consuming operation of obtaining of inverse stiffness matrix K^{-1} in solution of the system of equations (In.6) at every step of loading and at each iteration of the step, except the first iteration of the first step of loading (Fig. In.9).

Design diagram of structural analysis by finite element method for the allowance of the nonlinear properties according to the second way

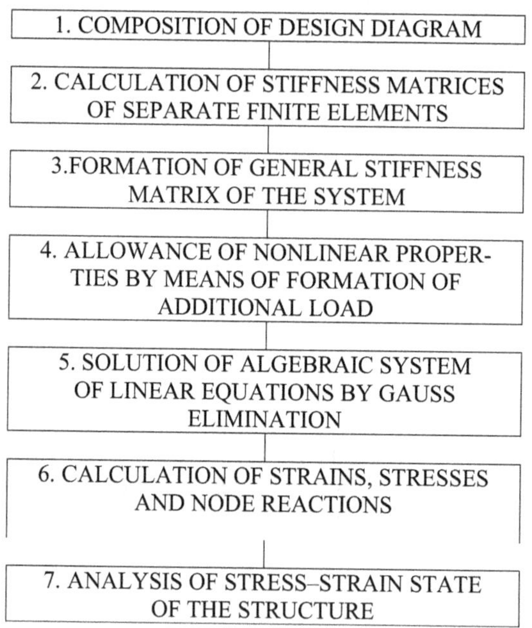

1. COMPOSITION OF DESIGN DIAGRAM
2. CALCULATION OF STIFFNESS MATRICES OF SEPARATE FINITE ELEMENTS
3.FORMATION OF GENERAL STIFFNESS MATRIX OF THE SYSTEM
4. ALLOWANCE OF NONLINEAR PROPERTIES BY MEANS OF FORMATION OF ADDITIONAL LOAD
5. SOLUTION OF ALGEBRAIC SYSTEM OF LINEAR EQUATIONS BY GAUSS ELIMINATION
6. CALCULATION OF STRAINS, STRESSES AND NODE REACTIONS
7. ANALYSIS OF STRESS–STRAIN STATE OF THE STRUCTURE

Fig. In.8.

Among the operations of direct execution of solution of system of algebraic linear equations only the operations connected with the calculation of constant terms are remained than at once to the operations of reverse execution are realized. Such approach to the construction of the process of iteration allows to reduce the labor consuming of the calculation essentially.

This approach leads to economy of machine time as in the solving of system of linear algebraic equations by Gauss elimination of unknowns approximately three quarters of time is occupied by the operation of obtaining of inverse stiffness matrix K^{-1}. And also this economy grows with increase of number of iterations at each step of analysis. Furthermore besides the economy of machine time the economy of required resource of computer for solving of the problem is reached.

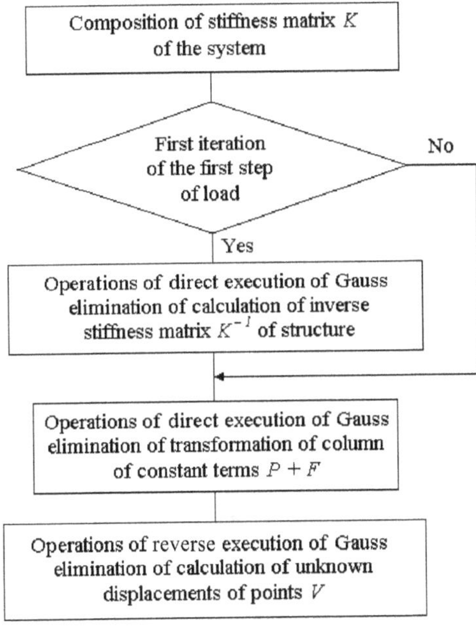

Fig. In.9. Flowchart of solution of algebraic system of linear equations $KV=P+F$

The main difficulty of use of the method of additional loads in nonlinear analysis of structures by finite element method is contained in obtaining of theoretical relationships which allow to form the vectors of additional load separately for each type of the considered nonlinearity and the vectors of correcting stresses in finite elements.

The simplest solving of this problem is the allowance of the plastic properties of material of the structure. It was proved by computations in practice. However if we take into account another types of nonlinearity, for example, cracking or reloading, the problem mentioned above becomes a very complicated one. The more so the necessary theoretical relationships ought to have the definite uniform to make possibility of simple algorithm for construction of computer programs.

Nevertheless, it should be noted that if the problem of formation of the vectors of additional load for the allowance of the different types of nonlinearity is solved, the programs constructed on base of this way for the allowance of the nonlinear properties are the more progressive ones due to their economy [8, 9, 34, 50, 67, 110].

In.3. Nonlinear analysis of reinforced concrete structures by finite element method

Analysis of reinforced concrete structures by finite element method is connected with necessity of allowance for strongly pronounced nonlinear properties. These nonlinear properties are connected with existence of various factors influencing the behavior of the structure under load. These factors have various natures and appear in various degrees at one or another stage of structural behavior.

In.3.1. Nonlinear properties of reinforced concrete structures

The main nonlinear properties of reinforced concrete structures are plasticity of concrete, bond between concrete and reinforcement, cracking, prestressing, reload action of temperature. The necessity of estimation of influence of these factors on behavior of the structure summons the necessity of development of the ways for the allowance of them in analysis by finite element method. Since the concrete has strongly pronounced plastic properties there were developed the ways for the allowance of them in analysis of reinforced concrete structures by finite element method. At the same time it should be noted that the majority of these ways is connected with change of stiffness matrix of the calculated structure, i.e. by action on the left hand side of the main decision system (In.3). The use of additional loads is more restricted. Whatever way for the allowance of the plastic properties by action on left or right side of the system (In.3) is used it requires the preliminary definition of means of the introductions of changes, i.e. assignment of the law of material deformation of the analyzed structure.

Recently the attention of many researches was attracted by study of diagram of concrete deformation connecting the relative deformations ε_b with the stresses σ_b in uniaxial compression and tension. The ways of obtaining of diagrams of concrete deformation are described explicitly in the work of N.I. Karpenko. There is a wide review of the works devoted to this problem in the same work too [49]. Nowadays it is considered that two-branched diagrams with descending part of branch of compression and tension are most close to description of real behavior of concrete of reinforced concrete structures. In particular the diagram of concrete deformation in compression with descending part of branch is described in detail in the work of Y.A. Ivashenko [46]. In analysis of structures the maximum values of tensile and compressive stresses of concrete are taken equal to the normative R_{btn} $(R_{bt,ser})$, R_{bn} $(R_{b,ser})$ or designed R_{bt}, R_b ones. Moreover the descending part of branch is not taken into account (Fig. In.10).

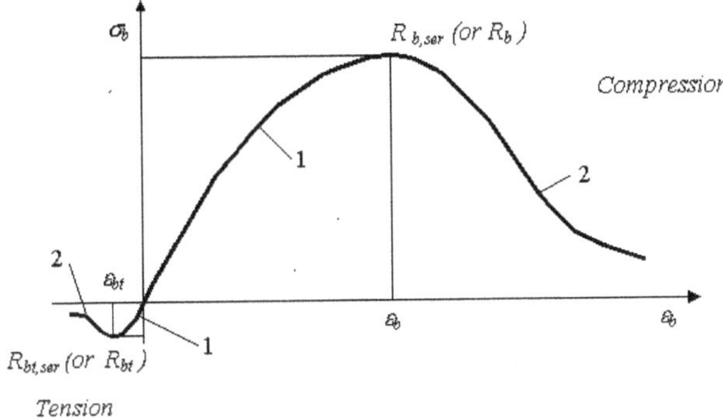

Fig. In.10. Two-branched diagram of concrete deformation in uniaxial compression and tension: 1 – ascending parts of branches, 2 – descending ones

Many factors such as conditions of load, action of temperature, cracking and so forth influence the form of the diagram of concrete deformation. Therefore the flexible easy changeable way for the allowance of the different variants of the used law of deformation is necessary for analysis of structures by finite element method. The way of elastic solution with introducing of additional loads meets such requirement completely. The main properties of triangular deep-beam finite element with plastic properties and the way of formation of additional load corresponding to finite element were determined by A.A. Karjakin [50, 86] on the basis of analogous finite element with linear properties and the theory of plasticity of G.A. Geniev [18] (Fig. In.11).

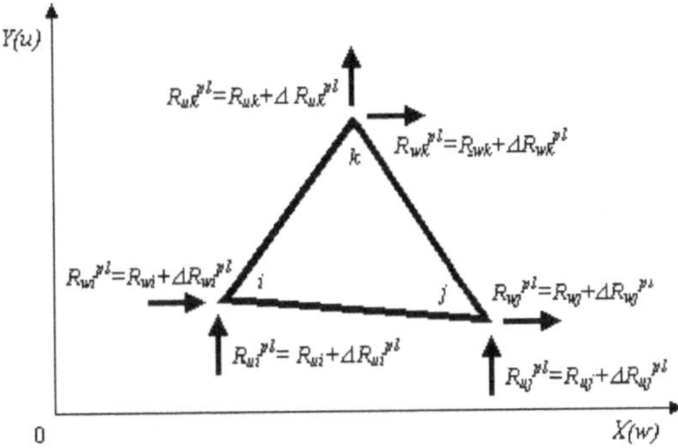

Fig. In.11. Scheme for the allowance of the plastic properties in finite element by means of additional load

This scheme shows how by means of the vector ΔR_{pl} the elastic node reactions R may be changed up to the value R_{pl}:

$$\Delta R_{pl} = tSC\Delta\sigma_{pl} \ , \qquad \text{(In.16)}$$

where $\Delta\sigma_{pl}$ = value characterizing the change of elastic stresses at expense of manifestation of plastic properties of concrete determined from the expression:

$$\sigma_{pl} = \sigma + \Delta\sigma_{pl}, \quad \text{(In.17)}$$

where σ = elastic stresses in finite element (see formula (In.7)); σ_{pl} = stresses of finite element for the allowance of the plastic properties of concrete.

The vector of additional load of the whole structure F_{pl} may be formed on the basis of the vectors ΔR_{pl} of separate finite elements. Efficiency of the use of this way for the allowance of the plastic properties of concrete was proved in the work [50] and confirmed in the works [25, 94, 95].

The analogous way of formation of additional load may be used for taking into account of other nonlinear singularities of behavior of reinforced concrete structures, for example, cracking. In a basis of this way the triangular deep-beam finite element with conditional crack may be used [25, 35, 53] (Fig. In.12).

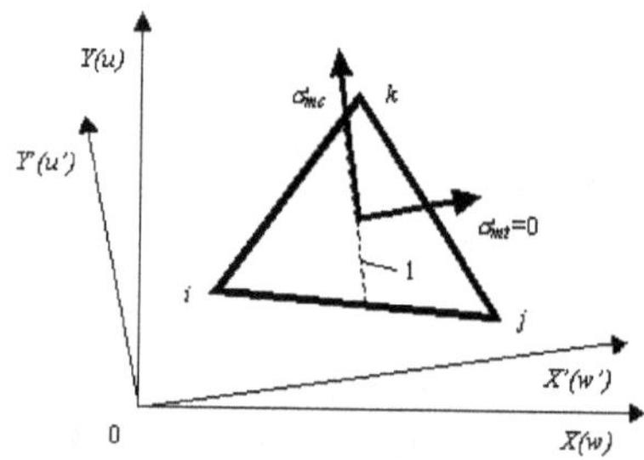

Fig. In.12. Triangular concrete deep-beam finite element with conditional crack.
1 – direction of cracking

In order to determine the properties of finite element with conditional crack it is necessary to clarify the degree of influence of crack on the stress-strain state of this finite element before the appearance of the crack. The next initial premises were accepted as preliminaries:

1. Cracks in concrete are formed when the main tensile stress σ_{mt} exceeds the limit resistance of concrete to tensile R_{bt}.

2. The direction of cracking is perpendicular to the direction of the main tensile stress σ_{mt}.

3. After appearance of crack the main tensile stress in the direction of perpendicular to the cracking becomes equal to 0, i.e. $\sigma_{mt} = 0$.

4. The centre of gravity of finite element is accepted as the place of appearance of crack.

The stresses σ_{crc} in this finite element are determined by the formula:

$$\sigma_{crc} = C_{crc}\sigma_{mc}; \quad \text{(In.18)}$$

where $C_{crc} = [\cos^2\alpha_1 \ \sin^2\alpha_1 \ -\sin\alpha_1\cos\alpha_1]$, α_1 = angle between the positive direction of axis X and direction of the main tensile stress σ_{mt} at the moment of appearance of crack, i.e. the direction of axis X'.

These stresses σ_{crc} may be presented as a result of action of two values:

$$\sigma_{crc} = \sigma + \Delta\sigma_{crc} \ , \qquad \text{(In.19)}$$

where σ = stresses in finite element without crack;

$\Delta\sigma_{crc}$ = value changing the stress σ up to the value σ_{crc} due to cracking.

After determination of the value $\Delta\sigma_{crc}$ the vector ΔR_{crc}, taking into account of cracking, may be obtained:

$$\Delta R_{crc} = tSC\Delta\sigma_{crc} \ . \qquad \text{(In.20)}$$

The principal scheme of taking into account of cracking in finite element by method of additional load is analogous to the scheme shown at Fig. In.11.

On the basis of vectors ΔR_{crc} of separate finite elements with crack the vector of additional load F_{crc} for the allowance of the cracking of the whole structure may be formed. Efficiency of this way for the allowance of the cracking was proved in the work [25].

Since there is no absolutely rigid bond between concrete and reinforcement these bonds are broken gradually and mutual displacements are developed at the surface of contact. Special connecting finite elements (CFE) are used to take into account this phenomenon.

The use of classical connecting finite element with elastic bonds (In.13) does not allow to take into account the nonlinear properties of the contact between concrete and reinforcement. At the same time this element allows to reflect the phenomenon by change of rigidity of its bonds in the process of calculation.

Behavior of the contact between concrete and reinforcement is described by differential law of bond between concrete and steel developed by A.A. Oatul as it is proved in the works [57, 66, 71–73, 102, 121]. Use of this law in analysis of reinforced concrete beams by finite element method in the work [50] as well as in analysis of the sample with centric axial tension in the work [25] gave a positive result.

On this basis of the properties of connecting finite element for simulation of the bond between concrete and reinforcement together with the corresponding method of formation of additional load (Fig. In.13) were developed.

where ΔK_{bond} = matrix turning the elastic stiffness matrix K of connecting finite element into nonlinear one K_{bond}.

On the basis of vectors ΔR_{bond} of separate connecting finite elements the vector of additional load F_{bond} of the whole structure is formed. Efficiency of this method for the allowance of the nonlinear properties of bond between concrete and reinforcement was proved in the work [25].

If reinforced concrete structure was subjected to a number of loadings with sequential unloading during the operating period its stress-strain state ought to be determined with regard of this phenomenon as it varies depending on the number of loading-unloading cycles [6, 7, 49, 69]. In connection of the fact it is necessary to take into account phenomenon in analysis of reinforced concrete structure by finite element method.

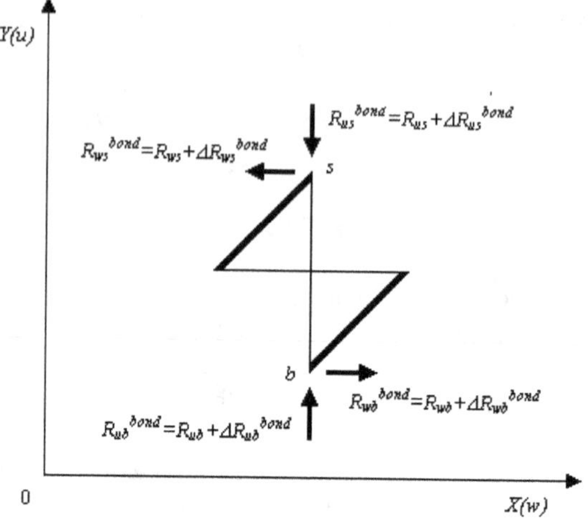

Fig. In.13. Scheme for the allowance of the nonlinear properties of bond between concrete and reinforcement in the connecting finite element by means of additional load

This scheme shows how by means of vector ΔR_{bond} to change elastic node reactions R up to the value of R_{bond}:

$$\Delta R_{bond} = \Delta K_{bond} V, \quad \text{(In.21)}$$

The diagrams of relationship between stresses and deformations for this case are represented at Fig. In.14. The form of diagram depends on either full or partial unloading after the first loading. The ways of construction of the diagrams are described in the work [49].

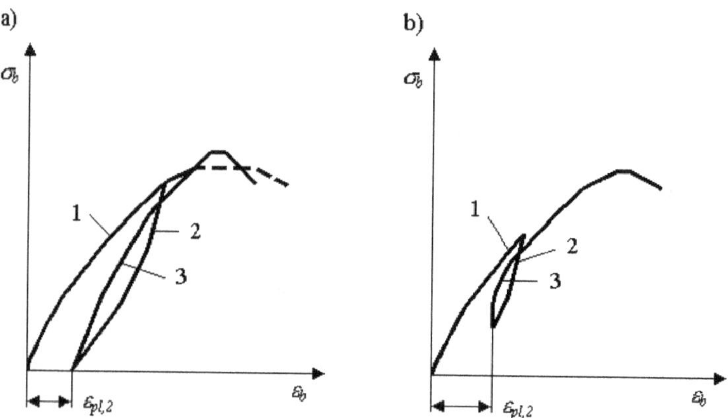

Fig. In.14. Diagrams of relationship between stresses and deformations of concrete under uniaxial compression in single reloading: a) full unloading; b) partial reloading; 1 – loading; 2 – unloading; 3 – reloading; σ_b = stresses of concrete; E_b = initial module of elasticity of concrete in loading; ε_b = deformation of concrete; $\varepsilon_{pl,2}$ = no recoverable plastic deformation of concrete in unloading or residual deformation

The main problem of concrete reloading is the estimation of residual deformations or stresses after initial loading and the determination of reloading diagram.

In analysis of reinforced concrete structures for the allowance of the temperature action by finite element method the algorithm of reduction of nonlinear temperature creeping defined by the temperature law by means of correcting coefficient to the law of concrete deformation (Fig. In.14) is used [84, 85] most often. Usually in addition the correction factors [49, 80] determined by empirical formulae at each step of analysis are used for the values E_b, R_{bt}, R_b.

The estimation of influence of prestressing on behavior of the structure is one of the problems in determination of real portrait of stress-strain state. The main point of prestressing is contained in creation of stresses opposite to the service load before the main load. Allowance of this phenomenon in analysis of reinforced concrete structures by finite element method represents a definite difficulty. According to a generally accepted approach the stress-strain state of a prestressed structure during its maintenance work under operating load may be presented as a result of the sequential action of the two main factors: prestressing and service load [90, 91].

At the same time in this case a lack of such simplified solution is that the influence of different nonlinear factors which are exhibited after prestressing before the service load are not taken into account. Such factors are: plastic properties of concrete, nonlinearity of bond between concrete and reinforcement, cracking.

Besides when the service load is applied a partial unloading due to prestressing takes place, i. e. the service load itself is a reload for the given structure. In analysis the introduction of prestressing as an external load the problem of its simulation in the design diagram of structure is appeared. In most cases the solution includes two problems: the determination of load's location and the way of its transmission. As to the location of prestressed force the problem is solved rather simple: the load is applied where the prestressed reinforcement is located. At the same time if we use finite element method a similar load may be applied only to the nodes of design diagram. This fact imposes definite requires for the way of transmission of this load depending on the considered structure and its design diagram [44]. In accordance with the division into finite elements two variants of relation between the size of finite elements along the prestressing line and the length of zone of prestressing transmission, i.e the anchorage length, are possible: this size may be more or less than the anchorage length. If the size of finite element is more then the anchorage length, the full value of prestressing is applied to a corresponding node of the design diagram. If the size of the finite element is less than the anchorage, the load of prestressing is applied to the several nodes in accordance with generally accepted linear diagram of its distribution. This variant is a slightly more complicated one then the previous variant but it is a very simple too. At the same time for the further analysis the procedure for the allowance of the nonlinear properties which appeared during prestressing in the analyzed structure is necessary.

In analysis of reinforced concrete structures their nonlinear properties are determined by plastic properties of concrete, cracking, prestressing and so on. Therefore the vector of additional load F at each stage of analysis ought to be determined according to the formula [25, 50]:

$$F = \sum_{i=1}^{n} F_i \; , \tag{In.22}$$

where F_i = vector of additional load taking into account of the character of appeared nonlinearity caused by one particular i-th factor; n = number of types of the nonlinearities which are taken into account at the given stage of structural analysis.

Table In.1 shows the example of formation of the vector of additional load F for reinforced concrete structures consisting of separate vectors F_i for the allowance of their more typical nonlinear properties [110].

The vector of additional loads F is calculated in order that its addition to the vector of external load P the structure with elastic properties should have the same displacements as this structure with linear properties.

In this case the values of obtained stresses ought to be corrected in the corresponding way. The essence of this process may be represented as the example of axial tension (compression) at Fig. In.15.

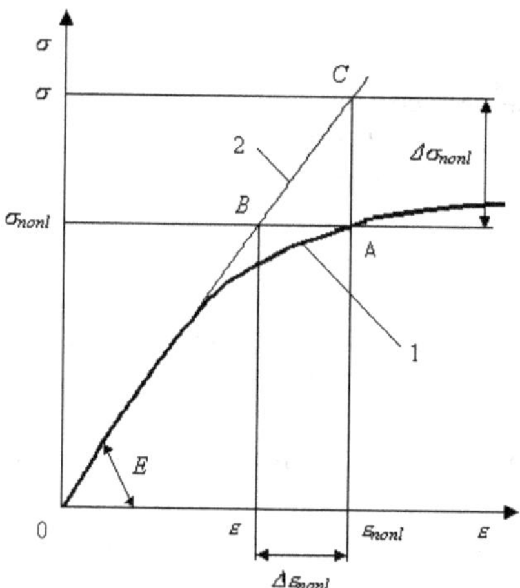

Fig. In.15. Diagram for the allowance of the nonlinear properties by means of the vector of additional load F: 1 – nonlinear relationship between strains ε_{nonl} and stresses σ_{nonl} (line OA); 2 – linear relationship between strains ε and stresses σ (line OBC); E = initial modulus of elasticity of material; ε_{nonl}, σ_{nonl} = nonlinear strains and stresses under external load P; σ= elastic stresses corresponding to nonlinear stresses ε_{nonl}, i.e. simultaneous applied external load P and additional load F; $\Delta\sigma_{nonl}$= value of change of elastic stresses σ up to σ_{nonl}; ε = elastic strains corresponding to nonlinear σ_{nonl}; $\Delta\varepsilon_{nonl}$= value of change of elastic strains ε up to nonlinear strains ε_{nonl}

As we can see the resulting stresses σ_{nonl} in the structure with nonlinear properties may be determined by the next formula:

$$\sigma_{nonl} = \sigma + \Delta\sigma_{nonl}\,, \qquad \text{(In.16)}$$

Table In.1

Formation of vector of additional load and correcting vector of stress for the allowance of the nonlinear properties of reinforced concrete structures

№	Type of nonlinearity	Considered nonlinearity	Vector of additional load		Correcting vector of stress	
			Notation	$F=\sum_{i=1}^{6} F_i$	Notation	$\Delta\sigma=\sum_{i=1}^{6}\Delta\sigma_i$
1	Plastic properties of concrete	1	F_{pl}	F_{pl}	$\Delta\sigma_{pl}$	$\Delta\sigma_{pl}$
2	Bond between concrete and reinforcement	1, 2	F_{bond}	$F_{pl}+F_{bond}$	$\Delta\sigma_{bond}$	$\Delta\sigma_{pl}+\Delta\sigma_{bond}$
3	Cracking	1, 2, 3	F_{crc}	$F_{pl}+F_{bond}+F_{crc}$	$\Delta\sigma_{crc}$	$\Delta\sigma_{pl}+\Delta\sigma_{bond}+\Delta\sigma_{crc}$
4	Prestressing	1, 2, 3, 4	F_{pr}	$F_{pl}+F_{bond}+F_{crc}+F_{pr}$	$\Delta\sigma_{pr}$	$\Delta\sigma_{pl}+\Delta\sigma_{bond}+$ $+\Delta\sigma_{crc}+\Delta\sigma_{pr}$
5	Repeated load	1, 2, 3, 4, 5	F_{rep}	$F_{pl}+F_{bond}+F_{crc}+F_{pr}+F_{rep}$	$\Delta\sigma_{rep}$	$\Delta\sigma_{pl}+\Delta\sigma_{bond}+$ $+\Delta\sigma_{crc}+\Delta\sigma_{pr}+\Delta\sigma_{rep}$
6	Temperature action	1, 2, 3, 4, 5, 6	F_t	$F_{pl}+F_{bond}+F_{crc}+F_{pr}+F_{rep}$ $+F_t$	$\Delta\sigma_t$	$\Delta\sigma_{pl}+\Delta\sigma_{bond}+\Delta\sigma_{crc}+$ $+\Delta\sigma_{pr}+\Delta\sigma_{rep}+\Delta\sigma_t$

where σ_{nonl} = stresses of the structure with linear properties;

σ = stresses of this structure with elastic properties, corresponding to simultaneous action of the external load P and the vector of additional load F, i.e. $(P+F)$;

$\Delta\sigma_{nonl}$ = correcting vector of stress for the allowance of the change of value of elastic stresses due to manifestation of nonlinear properties.

By analogy with formula (In. 15) for calculation of the vector of additional load F, the correcting vector of stresses $\Delta\sigma_{nonl}$ is determined for the allowance of all n types of nonlinearity considered at the given stage of calculation:

$$\Delta\sigma_{nonl} = \sum_{i=1}^{n}\Delta\sigma_i\,, \qquad \text{(In.24)}$$

where $\Delta\sigma_i$ = correcting vector of stresses for the allowance of the change of value of elastic stresses due to manifestation of definite type of nonlinearity.

Example of formation of the correcting vector of stress $\Delta\sigma_{nonl}$ from the separate vectors $\Delta\sigma_i$ for the allowance of the change of value of elastic stresses due to manifestation of the corresponding nonlinear properties of reinforced concrete structures is given in Table In.1.

As we can see in Table In.1 the algorithm of formation of the vectors of additional load F and the correcting vectors of stresses $\Delta\sigma_{nonl}$ for the allowance of the nonlinear properties is rather simple in mathematics and open to take into account of other types of nonlinearities that is important for the developed programs.

At the same time the creation of uniform algorithm is a serious problem due to variety of nonlinear properties of the structures. Particularly it concerns the reinforced concrete structures as it was discussed above.

In.3.2. Analysis of reinforced concrete structures at limit state and finite element method

At present the SN and P "Concrete and Reinforced Concrete Structures" [100, 101] are based on the use of limit state method (LSM). At the same time a serious problem is the combination of this method and the finite element method, which is oriented on the use of modern computers. Some aspects of the problem will be considered later.

The main statements of limit state analysis of reinforced concrete structures were developed by A.A. Gvozdev [17]. Later on the questions connecting with this method were studied by home and foreign scientists [2, 6, 11, 12, 15, 19–22, 42, 47, 54, 56, 58, 62, 70, 74, 77–79, 82, 83, 87, 93, 96, 100, 101, 108, 113, 114]. Now this method is widely used and stated in the SN and P.

The main point of limit state method is to determine the ultimate limit state of the structure and to guarantee its absence under conditions of the most unfavorable combination of loads and minimum characteristics of the strength of materials by introduction of the system of designed coefficients.

An ultimate limit state of the structure is the state in which the structure ceases to fulfill the function.

Reinforced concrete structures must be designed for two categories of limit states: the ultimate limit state and the serviceability limit state [90].

The loss of structural strength is the most dangerous state among the states of the first category. Most often design of the structure is reduced to the checking of its structural strength. The solution of this problem by finite element method is considered later.

At present the SN and Ps allow to approach to determination of stresses of the analyzed structures by two ways: with regard for nonlinear properties and without it in supposition of linear behavior. So section 1.15 of SN and P 2.03.01-84 "Concrete and Reinforced concrete Structures" stated:

"1.15 Stresses of the statically indeterminate reinforced concrete structures due to loads and forced displacements (in view of change of temperature, moisture content of concrete, displacement of supports and so on), and stresses of statically determinate structures in analysis according to the deformed scheme as a rule should be determined with regard for plastic stresses of concrete and reinforcement and in the presence of cracking.

For the structures without the developed procedure for the allowance of their inelastic properties as well as for the intermediate stages of analysis with regard for inelastic properties of reinforced concrete the stresses in the statically indeterminate structures are assumed to determine in the supposition of their linear elasticity."

This contradiction is reflected in a new version of SN and P 52-01-2003 and it appears especially explicit if the finite element method is used in structural analysis.

The recent approach consists of two steps of calculation: determination of stresses in members of structure by finite element method most often in supposition of their linear behavior and their design in accordance with limit state method.

For elimination of this contradiction it is necessary to solve the problem of obtaining of limit state for the considered structure by finite element method. It is possible only if nonlinear properties of the structure are manifested at the moment of ultimate limit state of the structure, i.e. with regard for the degree of influence and time of appearance of each separate nonlinear property. The problem is an actual one at present. The considered later an additional finite element method is one of the possible ways of its solution.

Chapter 1. BASIS OF METHOD OF ADDITIONAL LOADS FOR NONLINEAR ANALYSIS OF REINFORCED CONCRETE STRUCTURES AT LIMIT STATE

Realization of analysis of the reinforced concrete structures at limit state by finite element method (FEM) with allowance for nonlinear properties is impossible without the development of theoretical basis of a variant of this method destined for solution of the problem. Such variant is a suggested additional finite element method (AFEM) [41] as a combination of the three methods of structural design: finite element method [43, 123], method of additional loads [47] and limit state method [17].

1.1. Requirements for the developed method

On the basis of the study of additional loads use for nonlinear analysis of reinforced concrete structures by finite element method next requirements for the developed method of combination of finite element method, method of additional loads and limit state method may be stated:

1. Method ought to correspond to logic and sequence of the problem solving for definition of a stress-strain state for the designed structure by finite element method (FEM).
2. Method ought to have a generalized character independent of type, character and nature of structural nonlinear properties up to the moment when the ultimate limit state is reached, i.e. this method ought to be similar for all types of the considered nonlinearities, for example, for cracking as well as for temperature action.
3. Method in all its universality should be sufficiently flexible to reflect the essential singularities of each type of the considered nonlinear property in the process of ultimate limit state reaching, i.e. the result of its action in the case, for example for the allowance of the plastic properties ought to reflect the manifestation of these properties namely.
4. Method ought to reflect a change of strained state of designed structure due to the definite nonlinear property at the moment of reaching of ultimate limit state, i.e. the change of strains for example due to unload and reload ought to be characteristic just for this type of nonlinearity.
5. Method for the allowance of any nonlinear property ought to guarantee the possibility of existence of an accompanying method which takes into account stress-strain state change of designed structure due to manifestation at reaching of ultimate limit state, i.e. for example, allowance of the prestressing ought to be accompanied by the corresponding procedures of determination of the resulted strains of the structure.
6. The procedures of realization ought to be enough simple to allow a creation of algorithms and programs for nonlinear analysis of reinforced concrete structures at limit state on the basis of it.
7. The procedures ought to open a possibility for development of auxiliary algorithms and programs for transformation of programs for linear analysis of reinforced concrete structures into nonlinear analysis of the same structures at limit state.

It should be noted that enumerated requirements do not fulfill the equal functions. For example, the first requirement is a fundamental one; the second, third, fourth and fifth ones are the auxiliary requirements; the sixth and the seventh ones present consequences of computerization of the method.

1.2. Limit state of the structure and limit state of the finite element

It is known that the use of finite element method for analysis of any structure requires the definition of rigid properties of separate finite elements of its design diagram. In the main it concerns the construction of the stiffness matrix of each finite element, i.e. the definition of strains due to unit displacement of its nodes. These strains are determined by other methods of structural mechanics. Limit state method is the most effective one. Some questions connected with its use for nonlinear analysis of structural behavior at limit state by finite element method are considered in this section.

1.2.1. Limit state and an ideal failure model of the structure

An ultimate limit state of the structure is the state in which the structure ceases to fulfill the function.

Reinforced concrete structures must meet requirements for design accordance with two types of limit states: load capacity and serviceability [6].

The loss of structural strength is the most dangerous state among the states of the first type; therefore often the analysis of structure is reduced to checking of its strength. Precisely this problem of use of finite element method is considered later. It is proposed to use an ideal failure model for describing of structure at ultimate limit state which is the design diagram of an analyzed structure before the moment of its collapse.

The point is that as the structure is loaded its initial design diagram is changed in accordance with nonlinear properties which are manifested when the limit state is reaching. In particular the breakdown of bonds between some finite elements takes place, some finite elements with other nonlinear properties and the finite elements at limit state are appeared. As a result the initial design diagram is converted into ideal failure model of the considered structure.

This ideal failure model may be obtained by two ways. The first way is the carrying out analysis of the structure under step-by-step increasing load with introducing of accompanied changes in the initial design diagram. For example, the characteristics of the finite elements may be changed. The second way is the assignment of an ideal failure model known from design of analogous structures or from results of full-scale test. For example, SN and P for bending beams consider two failure models: with transverse crack and oblique one [90].

If we consider the concrete beam with bar reinforcement symmetrically loaded by concentrated forces we can see that its initial design diagram consists of two types of finite elements: the triangular deep-beam finite elements and the connecting finite elements (Fig. 1.1a, b). The first type is used for simulation of concrete and reinforcement. The second one is used for simulation of bond between concrete and bar reinforcement.

Since construction and load of the beam are symmetric its design diagram includes only a half of a beam up to the axis of symmetry. In this case the action of another part was compensated by horizontal bonds along the axis of symmetry. For simulation of limit state of beam it is necessary to insert the corresponding changes in its initial design diagram which respond to the character of failure. These changes determine the way of simulation of a dangerous crack.

Figures 1.1c and 1.1d represent the ideal failure models of the beam in the explicit form of its simulation by means of partition of nodes in the initial design diagram. Figures 1.2a and 1.2b show the ideal models of the cracks beam in the implicit form of the simulation by change of finite element properties, for example, by means of the corresponding additional finite elements (AFE).

When the first type of ideal failure model with transverse crack is used, it is accepted that dangerous transverse crack is formed in the middle of the beam along the axis of symmetry. In addition the limit height of compressive zone x above the transverse crack should be provided. Thus this ideal failure model in explicit form of simulation differs from the initial design diagram by absence of some horizontal bonds of concrete finite elements along the axis of symmetry. These horizontal bonds are only on the height of the compressive zone x of section (Fig. 1.1c).

In implicit form of crack simulation (Fig. 1.2a) the properties of finite elements along the axis of symmetry under the height of compressive zone of section x should be changed by means of the corresponding additional finite elements.

In the second case of failure with oblique crack another model should be used. In explicit form of simulation of oblique crack the design diagram differs from the initial one by presence of breakdown between concrete finite elements along the direction of crack with plane view c_0 from support, as well as by introducing of additional connecting finite elements at the points of intersection of crack with reinforcement. In addition the height of compressive zone x above the crack should be provided too (Fig. 1.1d). In implicit form of simulation of oblique crack (Fig. 1.2b) the properties of finite elements situated along the direction of oblique crack should be changed. These changes may be done by the corresponding additional finite elements.

In the presence of two ideal failure models of the bending beams it is necessary to carry out analysis of the structure according to each one.

Introducing of ideal failure model opens the wide opportunity for analysis of reinforced concrete structures at limit states by finite element method (FEM) [38, 40] and allows to develop the additional finite element method (AFEM).

1.2.2. Limit state of the finite element

There is its own stress-strain state at each point of the structure during operating under load. The failure begins in the point with ultimate limit state. The degree of inclusion of each point in general behavior of the structure is determined by the degree of reaching of ultimate limit state in it. The separate finite element is considered instead of the point in analysis of structure by finite element method.

The structure is considered as an assembly of such elements where each finite element is the separate small structure of simple form fixed in nodes. This finite element has its own limit state in operating under load. The contribution of each finite element in behavior of the structure is determined by the degree of limit state reaching in it too. It means that rigid properties of the main finite element introducing in analysis ought to be formed in connection with the degree of limit state reaching stage in all its points.

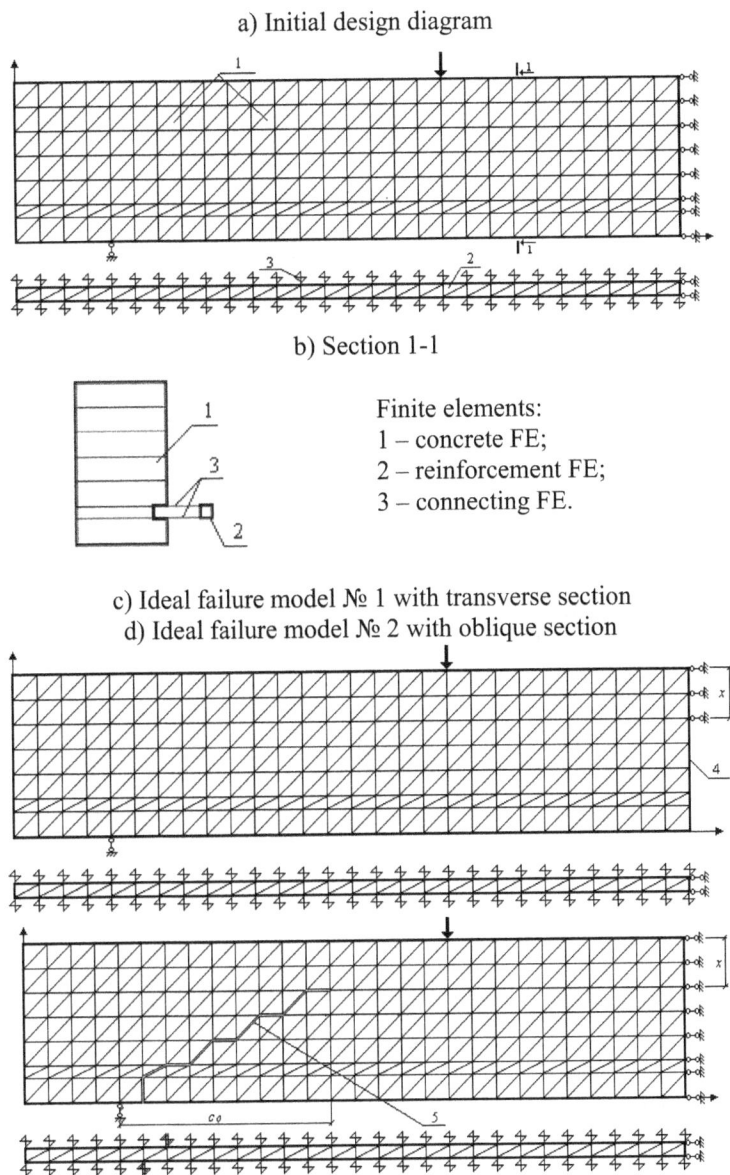

a) Initial design diagram

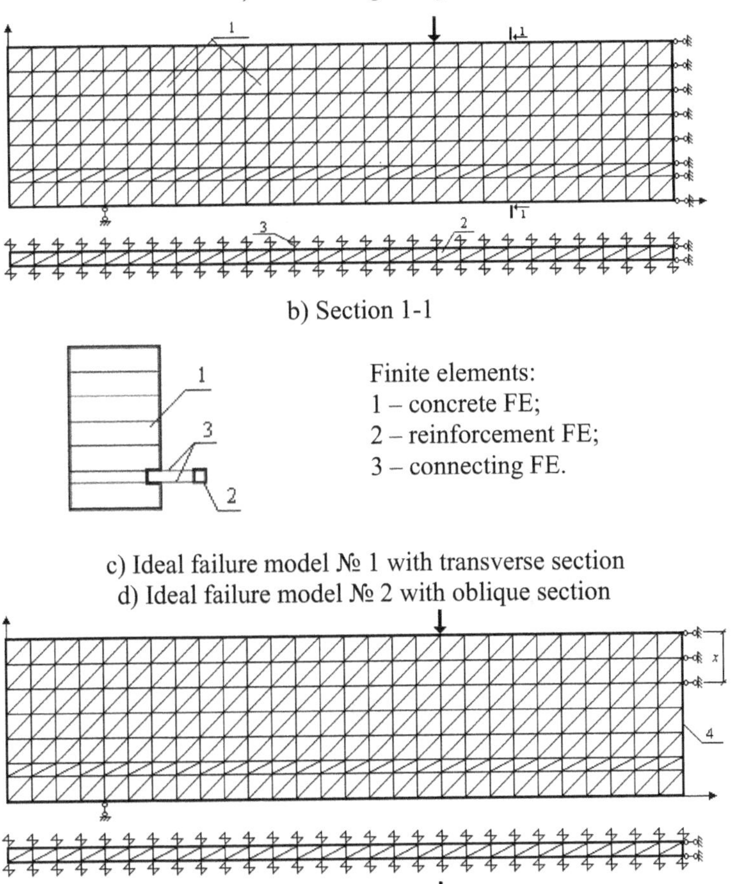

Fig. 1.1. Ideal failure models of bending beam for simulation of crack in explicit form: 4 – transverse crack; 5 – oblique crack; x – height of compressive zone; c_0 – plane view of the oblique crack.

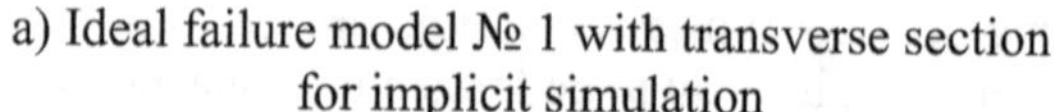
a) Ideal failure model № 1 with transverse section for implicit simulation

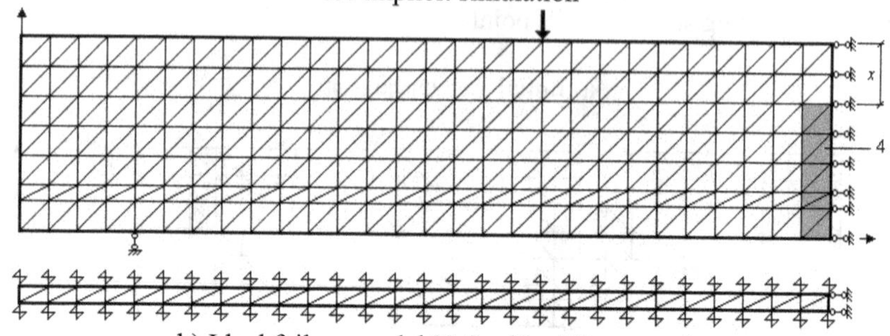

b) Ideal failure model № 2 with oblique section for implicit simulation

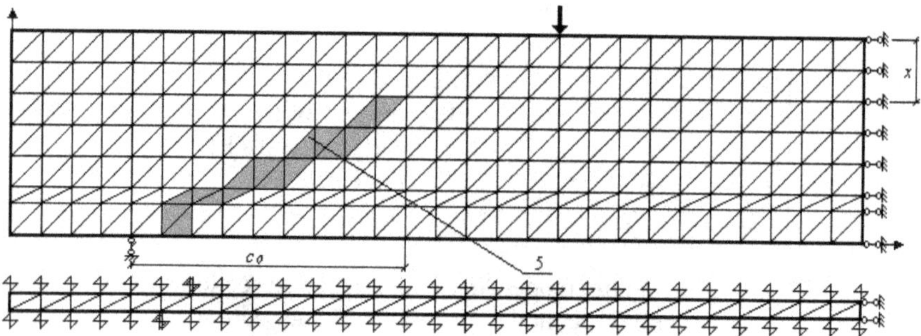

Fig. 1.2. Ideal failure model of bending beams for simulation of crack in implicit form: 4 – finite elements, which properties should be changed for simulation of transverse crack; 5 – finite elements, which properties should be changed for simulation of oblique crack; x – height of compressive zone; c_0 – plane view of oblique crack

It should be noted that a special research of each finite element of definite type is necessary to clarify the number of its typical limit states because it may have several ones. Besides in solving of nonlinear problem this research ought to determine typical for the given finite element nonlinear properties, the moment of their appearance and the degree of influence on a stress-strain state at each moment of its behavior from the beginning to the end.

Necessity of such research is especially important in case when the ultimate limit state of the structure is reached after limit state in the only finite element, for example, for the bar systems. The main input data for such research are next factors: form and material of finite element, stress-strain state in the stage of ultimate equilibrium and after it. Next two factors may be the key factors of the end of the ultimate limit states:

a) reaching of ultimate limit state if after it the finite element ceases to fulfill the function;

b) definition of character of the finite element behavior if after the ultimate limit state it partly continues to take part in general behavior of the structure.

For example, since concrete has two forms of failure (due to tension and compression), then there are at least two limit states for each finite element. Moreover if the finite element ceases its existence as a member of the structure after the failure due to compression the fact is that after the failure due to tension it may be partly included in operating under compression.

It should be noted that in general the moments of reaching of ultimate limit state in the separate finite element and in the whole structure do not coincide. It is explained by the fact that the reaching of ultimate limit state in one point, i.e. in the separate finite element is far from being always an ultimate limit state of the whole structure. The ultimate limit state of the structure appears if the ultimate limit state is reached in a number of finite elements. The definition of these finite elements is the aim of analysis of behavior of the structure by finite element method. The approach to solving of nonlinear problem of analysis of reinforced concrete structures by finite element method for the allowance of the limit state of separate finite elements appears rather promising although its realization is connected with the serious difficulties.

1.2.3. Limit states of the triangular concrete deep-beam finite element

The definition of concept of the ultimate limit state of a separate finite element and the necessity to research the process of its reaching were considered in the previous section [36, 112]. The research of behavior of triangular concrete deep-beam finite element may be an example of such work. The properties of this finite element are considered in In. 1.3 and In. 3.1. Two main stresses σ_{mc} and σ_{mt} appear in behavior of such finite element. The destruction of concrete may occur by two ways: if the main compressive stress σ_{mc} exceeds the limit compressive strength R_b ($\sigma_{mc} > R_b$) and if the main tensile strength σ_{mt} exceeds the limit tensile strength R_{bt} ($\sigma_{mt} > R_{bt}$). Therefore there are two limit strength states of triangular concrete deep-beam finite element: limit compressive strength and limit tensile strength (Fig. 1.3). The limit compressive strength has two variants of behavior of finite element in plane stress state (Fig. 1.3a):

1) limit compression in the direction of stress $\sigma_{mc} > R_b$ and compression in the direction of another stress $\sigma_{mt} < R_b$;

2) limit compression in the direction of stress $\sigma_{mc} = R_b$ and tensile in the direction of another stress $\sigma_{mt} < R_{bt}$.

These two cases may be combined, i.e. after exceeding by stress σ_{mc} of the limit value which is equal to limit compressive strength R_b , ($\sigma_{mc} > R_b$) the failure of the finite element takes place and concrete ceases to fulfill the function.

The limit tensile strength has two variants of behavior of finite element (Fig. 1.3b):

1) limit tension in the direction of stress $\sigma_{mt} = R_{bt}$ and compression in the direction of another stress $\sigma_{mc} < R_b$;

2) limit tension in the direction of stress $\sigma_{mt} = R_{bt}$ and tension in the direction of another stress $\sigma_{mc} < R_{bt}$.

These two cases may be combined too, i.e. after exceeding by the stress σ_{mt} of the limit value which is equal to the limit tensile strength of concrete R_{bt} ($\sigma_{mt} > R_{bt}$) the crack takes place in concrete. In addition concrete remains its bearing capacity partly in the direction of another stress σ_{mc}. Such approach reflects the fact that SN and Ps allow operating of reinforced concrete structures with cracks [90, 91]. Now we may study the behavior of this finite element in each out of two ultimate limit states. Under compression before and after reaching of ultimate limit state the triangular deep-beam finite element passes through several states which reflect the singularities of behavior of concrete as a structural material:

1) initial load of the finite element, i.e. an inclusion of the finite element in behavior of the whole structure;

2) plastic behavior of the finite element in plane stress state, i.e. when $\sigma_{mc} < R_b$ and $\sigma_{mt} < R_{bt}$, if σ_{mt} is a tensile stress or $\sigma_{mt} < R_b$, if σ_{mt} is a compressive one;

3) behavior at ultimate limit state of compression, i.e. if $\sigma_{mc} = R_b$;

4) exceeding by the main compressive stress σ_{mc} of limit compressive strength of concrete R_b and failure of the finite element;

5) elimination of the finite element out of behavior of the structure as after the failure due to compression concrete does not exist as a structural material and it loses the bearing capacity completely.

In general if we say of behavior of triangular finite element at ultimate compression two main stages may be distinguished: 1) plastic behavior at plane stress state before the moment of reaching of ultimate limit state in compression; 2) complete failure after reaching of this limit state. In the considered example the moment of reaching by the main compressive stress σ_{mc} of the limit compressive strength R_b ($\sigma_{mc} = R_b$) was taken as a criterion of ultimate limit state. If it is necessary another factor may be accepted as a criterion, for example, reaching by concrete of the limit strain in compression. The choice of criterion depends on the conditions of analysis.

In the second case of behavior of finite element in tension after reaching of ultimate limit state triangular deep-beam finite element is eliminated partly since it may perceive some stresses (see In.3.1). It means that the finite element passes through next stages of behavior:

1) initial load of the finite element, i.e. an inclusion of finite element in behavior of the whole structure;

2) plastic behavior of the finite element in plane stress state, i.e. when $\sigma_{mt} < R_{bt}$ and $\sigma_{mc} < R_{bt}$, if σ_{mc} is a tensile stress or $\sigma_{mc} < R_b$, if σ_{mc} is a compressive one;

3) behavior at ultimate limit state of tension, i.e. if $\sigma_{mt} = R_{bt}$;

4) exceeding by the main tensile stress σ_{mt} of limit tensile strength of concrete R_{bt}, cracking and partly reload of the finite element;

5) repeated partial inclusion of the finite element in behavior of the whole structure in the direction of the main stress σ_{mc};

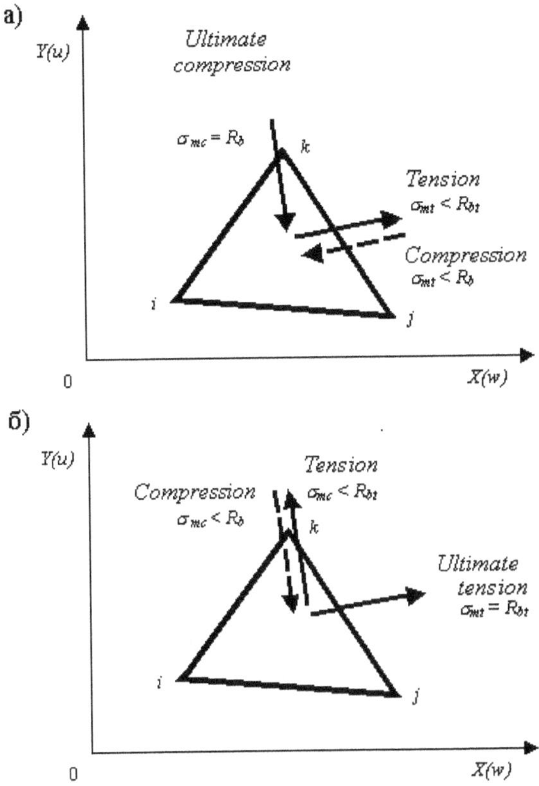

Fig. 1.3. Ultimate limit strength state of triangular concrete deep-beam finite element: a) ultimate compression; b) ultimate tension; σ_{mc} = main compressive stress; R_b = limit compressive strength of concrete; σ_{mt} = main tensile stress; R_{bt} = limit tensile strength of concrete

6) behavior at ultimate limit state of another main stress, i.e. if $\sigma_{mc}= R_b$ or $\sigma_{mc}= R_{bt}$;

7) exceeding by the main stress σ_{mc} of the limit compressive strength of concrete R_b or the limit tensile strength of concrete R_{bt} and a failure of the finite element;

8) elimination of the finite element out of behavior of the structure as concrete loses its bearing capacity completely.

Thus in behavior of the triangular finite element at ultimate limit state of tension we may say of next four stages:

1) plastic behavior of the finite element at plane stress state up to the moment of reaching of ultimate state of tension;

2) reload in part of the finite element after the reaching of ultimate limit state due to cracking;

3) partial plastic behavior up to the moment of ultimate limit state reaching at reloading;

4) collapse after reaching of this ultimate limit state.

In this case the moment of reaching by the main tensile stress σ_{mt} of limit strength in tension R_{bt} ($\sigma_{mt} = R_{bt}$) was accepted as the first criterion which leads to loss of its partial bearing capacity. The moment of reaching by the main tensile stress σ_{mt} of limit strength in compression of concrete R_b, if this stress σ_{mt} is compressive or of limit strength in tension R_{bt}, if σ_{mt} is a tensile one was accepted as the second criterion which leads to complete loss of bearing capacity.

The choice of these two criterions depends on the conditions of the carrying out analysis. For example, if it is necessary the limit strain in tension may be accepted as the first criterion and the limit strain in compression may be accepted as the second one or the limit strain in tension may be accepted as the second criterion too.

The limit states of the triangular concrete deep-beam finite element were considered in this section but these states are typical for concrete deep-beam finite element of another geometric shape, for example, a rectangular one. It stems from the general behavior of concrete at plane stress state.

1.2.4. Limit states of the bar of general attitude

"The bar of general attitude" is often used for simulation of spatial frames. The description of its system of stresses was given in section 1.1.5 [64]. Each stress has its own limit state which is determined by the properties of a finite element material, concrete in this case.

Concrete is characterized by two types of strength (compressive and tensile). It is a determining fact for limit states of this finite element. Table 1.1 illustrates this process and reflects the peculiarities of behavior of concrete in spatial frames [40].

Behavior of the concrete bar finite element at each limit state mentioned above demands a special theoretical research.

In particular, since we may come across the proper shear in reinforced concrete rarely, the appearance of shear limit state is low-probability (see Table 1.1, sections 5.2 and 6.2).

The results of such research are very important for analysis of the bar systems if ultimate limit state of the whole structure is determined by the moment of reaching of ultimate limit state at one of its finite elements, i.e. by the moment of its failure. It occurs rather frequently.

Introduction of the concepts of ideal failure model of the structure and ultimate limit state of separate finite element allows combining such effective means of analysis as finite element method and limit state method which today are the most effective methods of analysis of the building structures.

Table 1.1

Limit states of concrete finite element «The bar of general attitude»

№	FORCE	LIMIT STATES	CHARACTER OF FAILURE
1	N – longitudinal force	1. Tensile stress reaches the value of tensile strength of concrete	Tension
		2. Compressive stress reaches the value of compressive strength	Compression
		3. Longitudinal compressive force reaches the ultimate value in static stability design	Lost of stability
2	M_y – bending moment about axis *y*	1. Tensile bent stress in *y* direction reaches the value of tensile strength of concrete	Failure of tensile zone of the border of section
		2. Compressive bent stress in *y* direction reaches the value of compressive strength of concrete	Failure of compressive zone of the border of section
3	M_z – bending moment about axis *z*	1. Tensile bent stress in *y* direction reaches the value of tensile strength of concrete	Failure of tensile zone of perpendicular border of section
		2. Compressive bent stress in *z* direction reaches the value of tensile strength of concrete	Failure of compressive zone of perpendicular border of section
4	M_k – torsion moment about axis *x*	1. Torsion moment reaches the ultimate value in torsion with bent	Failure due to torsion with bending
5	Q_y – shear force in *y* direction	1. Shear force in *y* direction reaches the ultimate value in strength of oblique section in *xoy* plane	Failure of oblique section in *xoy* plane
		2. Tangential stress in *xoy* plane reaches the shear strength of concrete	Failure due to shear in *xoy* plane
6	Q_z – shear force in *z* direction	1. Shear force in z direction reaches the ultimate value in strength due to oblique section in *xoz* plane	Failure of oblique section in *xoz* plane
		2. Tangential stress reaches the ultimate value of shear strength of concrete in *xoz* plane	Failure due to shear in *xoz* plane

1.3. The additional design diagram

The concept of ideal failure model as a design diagram at limit state was stated in section 1.2.1.

The development of gradual passage from the initial design diagram to the ideal failure model is the next step. It is necessary to introduce the concepts of additional design diagram and additional finite element [36, 39, 112].

1.3.1. Function of additional design diagram

Starting from the first main requirement for the developed technique the decision system of linear algebraic equations (In.2) at limit state ought to be of the next form:

$$K_{lim} V = P\,, \tag{1.1}$$

where V = matrix-column of unknown node displacements;

P = matrix-column of outer load;

K_{lim} = stiffness matrix of the structure with nonlinear properties at limit state.

This nonlinear matrix K_{lim} is formed on the basis of design diagram of the structure and represents its ideal failure model, i.e. an assembly of separate finite elements where each finite element has its own nonlinear properties in connection with the degree of its reaching limit state. If the way of elastic decision, which is the basis of additional loads, is used matrix K_{lim} ought to be in the form (In.13), i.e. the linear and nonlinear components ought to be distinguished. It is convenient to repeat the writing [81]:

$$K_{lim} = K + \Delta K_{lim}\,, \tag{1.2}$$

where K = initial stiffness matrix of the structure with linear properties;

ΔK_{lim} = matrix connected with manifestation of nonlinear properties of the structure at limit state.

Since nonlinear matrix ΔK_{lim} is permanently changed in the process of analysis due to appearance of different nonlinear properties in connection with the degree of loading of the structure then the isolation of its linear and nonlinear components K and ΔK_{lim} is the process of definite difficulty.

More simply it is to use as linear component K the stiffness matrix of the structure if it is made on the basis of the same design diagram and consists of finite elements with linear properties. It permits to leave the linear matrix K as a constant part from the beginning to the end of analysis.

Now let us pass to the nonlinear term ΔK_{lim}, which may be determined from the formula (1.2):

$$\Delta K_{lim} = K_{lim} - K\,. \tag{1.3}$$

In the formula (1.3) both matrices K_{lim} and K are formed on the basis of one and the same design diagram, which consists of identical finite elements with equal dimensions and shape.

But the first nonlinear matrix K_{lim} consists of the finite elements with nonlinear properties in an accordance with the degree of reaching by these finite elements of limit states. The second linear matrix K consists of the same finite elements but with linear properties. It means that nonlinear component ΔK_{lim} is an additional matrix which ought to be formed on the basis of the same design diagram. It ought to be added to linear matrix K for obtaining of stiffness matrix for the allowance of the nonlinear properties ΔK_{lim}. It should be formed on the basis of the same finite elements by means of which linear matrix K and nonlinear matrix K_{lim} are formed. We may assume that additional matrix ΔK_{lim} is formed on the basis of an additional design diagram of the structure which consists of the additional finite elements.

This additional design diagram allows to turn design diagram consisting of linear finite elements into design diagram with nonlinear finite elements as each its additional finite element converts the corresponding linear finite element into nonlinear one. The additional design diagram is used for formation of additional load which takes into account limit state of the structure [37]. With regard formula (1.2) the main system (In.4) takes the form:

$$KV = P - \Delta K_{lim} V . \tag{1.4}$$

The second term of the right-hand side of this equation is the additional load F which ought to be applied to linear system together with the main load P to reach the displacements of corresponding nonlinear system at limit state under the action of the load P only:

$$F = -\Delta K_{lim} V . \tag{1.5}$$

The types of actions of additional design diagram consisting of additional finite elements are given in Fig. 1.4. The information about additional finite elements will be given later.

a) The transformation from initial design diagram to design diagram of structure at limit state

Ideal failure model – design diagram of structure at limit state	=	**Initial design diagram consisting of finite elements with linear properties**	+	**Additional design diagram for the allowance of the limit state of the structure**

b) The transformation from design diagram of structure at limit state to initial design diagram consisting of finite elements with linear properties

Initial design diagram consisting of finite elements with linear properties	=	**Ideal failure model – design diagram of structure at limit state**	–	**Additional design diagram for the allowance of the limit state of the structure**

c) The formation of additional loads for the allowance of the limit state of the structure

Additional design diagram for the allowance of the limit state of the structure	=	**Ideal failure model – design diagram of structure at limit state**	–	**Initial design diagram consisting of finite elements with linear properties**

Fig. 1.4. Types of action of additional design diagram for the allowance of the structural limit state

1.3.2. Formation of the additional design diagram

Additional stiffness matrix of the structure ΔK_{lim} is formed on the basis of nonlinear properties of the finite elements of its design diagram as the structure at limit state exhibits all intrinsic nonlinear properties. These nonlinear properties have different nature and character of manifestation. Therefore the additional stiffness matrix ought to be formed in accordance with nonlinearities observed at the moment of limit state:

$$\Delta K_{lim} = \sum_{i=1}^{n} \Delta K_{nonl,i} , \tag{1.6}$$

where n = number of types of nonlinear properties of the structure at limit state;

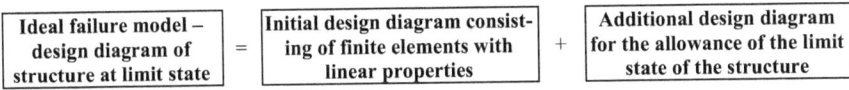

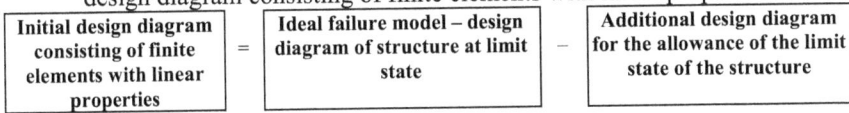

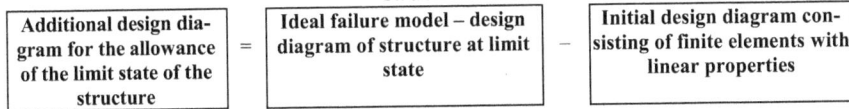

$\Delta K_{nonl,i}$ = additional stiffness matrix of the structure for the allowance of the manifestation of its i-th nonlinear properties at the moment of reaching of limit state. This additional stiffness matrix of the structure $\Delta K_{nonl,i}$ determines the degree of influence of the i-th nonlinear properties on behavior of structure at limit state. In view of this the stiffness matrix is determined according the next formula:

$$\Delta K_{nonl,i} = K_{nonl,i} - K_{nonl,i-1} , \qquad (1.7)$$

where $K_{nonl,i}$ = stiffness matrix of structure at limit state with regard for the i-th considered nonlinear property;

$K_{nonl,i-1}$ = stiffness matrix of structure at limit state without regard for the i-th nonlinear property.

In general in the presence of the given nonlinear properties in the moment of reaching of ultimate limit state the additional matrix $\Delta K_{nonl,i}$ is not equal to 0, i.e.

$$\Delta K_{nonl,i} \neq 0 . \qquad (1.8)$$

If the i-th nonlinear property at ultimate limit state is not taken into account its additional matrix $\Delta K_{nonl,i}$ is equal to 0, i.e.

$$\Delta K_{nonl,i} = 0 . \qquad (1.9)$$

With regard to the first nonlinear property the stiffness matrix of structure $\Delta K_{nonl,1}$ may be determined by the next formula:

$$\Delta K_{nonl,1} = K_{nonl,1} - K_{nonl,0} = K_{nonl,1} - K , \qquad (1.10)$$

where K = stiffness matrix of structure which is formed on the basis of the finite elements with linear properties.

In case of analysis without allowance of the nonlinear properties the additional stiffness matrix ΔK_{lim} is not taken into account and is equal to 0:

$$\Delta K_{lim} = 0 . \qquad (1.11)$$

Each finite element of the matrix is equal to 0 too (see (1.9)).

The additional stiffness matrix of structure ΔK_{lim} is formed on the basis of additional design diagram. It means that additional design diagram ought to be determined in accordance with the number n of nonlinear properties at limit state, i.e. it ought to consist of n additional design diagrams where each design diagram considers the i-th nonlinear property only. If any nonlinear property is not considered at the moment of limit state reaching then the corresponding additional design diagram is absent too.

The general scheme of formation of additional design diagram for the allowance of the limit state of the structure is given in Fig. 1.5.

Example of its formation in analysis of reinforced concrete structures at reaching of limit state in accordance with the most intense manifestation of nonlinear properties is given in Fig. 1.6. In this case six nonlinear properties ($n = 6$) enumerated in Table In.1 were considered:

1) plastic properties of concrete;
2) nonlinearity of bond between concrete and reinforcement;
3) cracking;
4) prestressing;
5) reload and unload;
6) action of temperature.

Stated above approach to the formation of additional design diagram of structure does not introduce a restriction of the considered nonlinear properties in type, character and number of any structures at limit state (reinforced concrete structures among them) in analysis by finite element method.

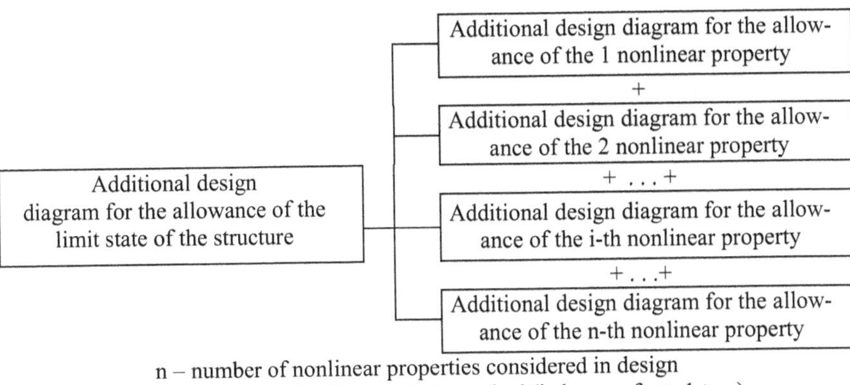

Fig. 1.5. Additional design diagram for the allowance for the limit state of the structure

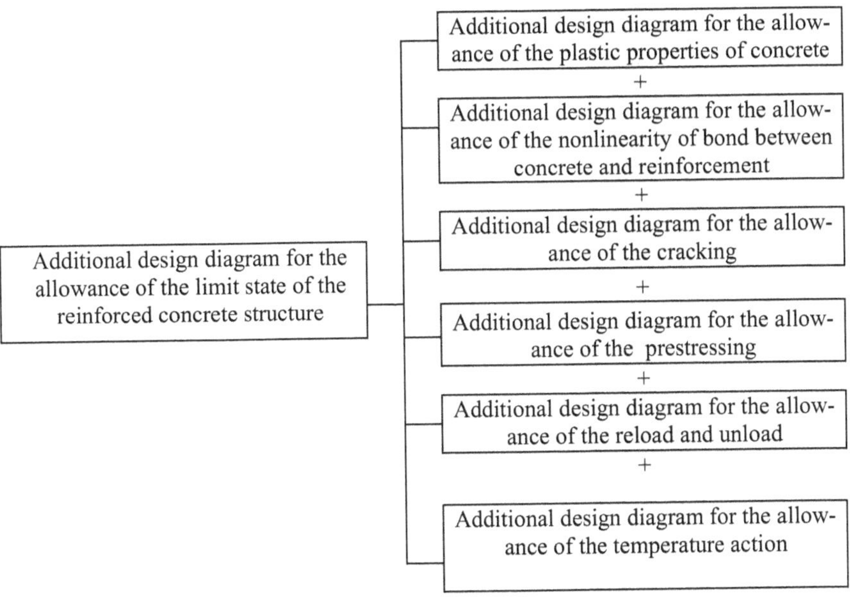

Fig. 1.6. Additional design diagram for the allowance of the nonlinear properties (n=6) of the reinforced concrete structure at limit state

Such principle of determination of the additional design diagram properties opens the broad opportunities for analysis of structure at limit state by finite element method as it admits a gradual change of properties of initial design diagram consisting of finite elements with linear properties.

Formation of additional design diagram in the case if an initial design diagram consists of the triangular concrete deep-beam finite elements may be mentioned as a specific example.

Starting from the main stresses appeared in triangular concrete finite element there are two types of analogous limit states in it (see section 1.2.3). In this connection the behavior of finite element has different stages. Thus in compression the finite element has two stages which reflect the singularity of concrete as a material. These are:

1) plastic behavior up to the moment when the limit state is reached;

2) collapse of the finite element after the compressive failure of concrete as at this moment concrete loses its bearing capacity and ceases existing as integral material.

Under tension when the limit state is reached the finite element loses its bearing capacity in part as well as it may have some stresses. This means that the finite element has four stages up to the moment of collapse:

1) plastic behavior before cracking;

2) cracking and unloading in part;

3) partly reload up to the moment when the limit state of reloading is reached;

4) collapse.

Since the additional design diagram ought to take into account the character of behavior of each finite element of the initial design diagram the order of its formation is determined by all stages of behavior of the main element at each limit state by means of the corresponding additional finite elements (Fig. 1.7).

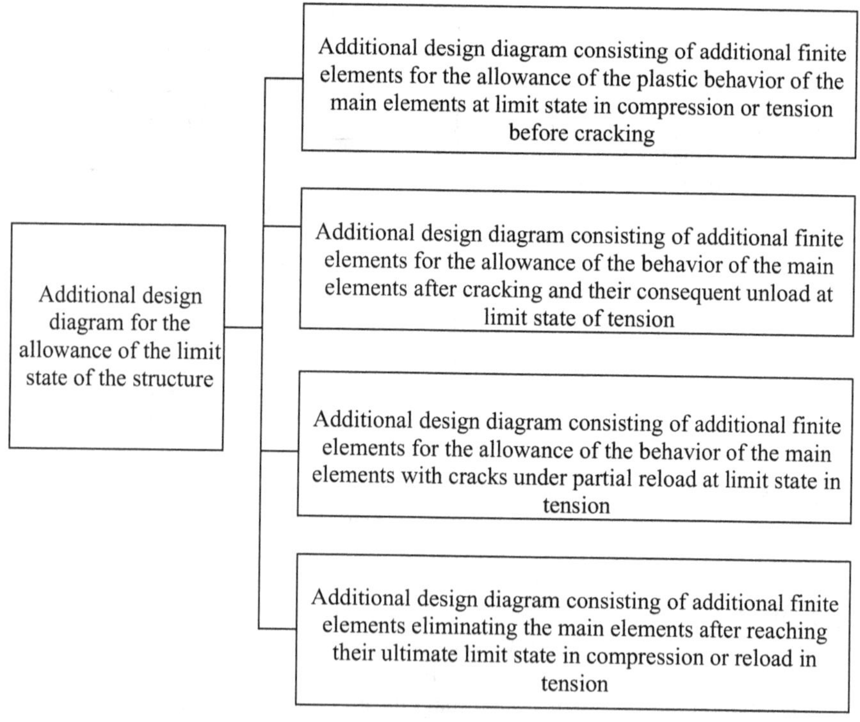

Fig. 1.7. Example of formation of additional design diagram consisting of additional finite elements

The considered example of use of additional triangularar concrete deep-beam finite elements determines the way of realization of limit state analysis of reinforced concrete structures at plane stress state.

1.4. Additional finite elements

Additional design diagram for the allowance of the nonlinear properties of the considered structure at limit state ought to be constructed on the basis of special additional finite elements each of which introduces in analysis of nonlinear properties of the separate finite element as its limit state is reached.

1.4.1. Function of additional finite elements

It was proved in the previous section that additional design diagram ought to be formed on the basis of the initial design diagram. It shows the analysis of the formula (1.3) for determination of its stiffness matrix ΔK_{lim}.

Linear stiffness matrix K in this formula is formed on the basis of finite elements with linear properties and matrix K_{lim} is formed on the basis of the same finite elements but with nonlinear properties. Linear finite elements and nonlinear ones have equal shape and dimensions but different stiffness characteristics.

Thus the finite elements of the same shape and dimensions ought to be the base of additional design diagram. These finite elements are additional finite elements.

In this case additional stiffness matrix is a combination of the additional finite elements and it is able to turn design diagram consisting of linear finite elements into design diagram consisting of nonlinear finite elements.

In turn each additional finite element transforms a finite element with linear properties into a finite element with nonlinear properties depending on the stage of behavior of the main finite element on the way to its ultimate limit state.

To illustrate the above said let us write down the next formula for determination of stiffness matrix of finite element with nonlinear at the moment of reaching of ultimate limit state of the structure:

$$K_{nonl,e} = K_e + \Delta K_{nonl,e} \ , \qquad (1.12)$$

where $K_{nonl,e}$ = stiffness matrix of finite element with nonlinear properties at the moment of reaching of ultimate limit state of the structure;

K_e = stiffness matrix of finite element with linear properties;

$\Delta K_{nonl,e}$ = stiffness matrix of additional finite element which allows to transform the stiffness matrix of finite element with linear properties into the stiffness matrix of finite element with nonlinear properties.

This matrix $\Delta K_{nonl,e}$ may be named an additional stiffness matrix of the considered finite element and it may be determined by the next formula:

$$\Delta K_{nonl,e} = K_{nonl,e} - K_e \ . \qquad (1.13)$$

In general, i.e. in analysis with allowance for nonlinear properties this additional stiffness matrix $\Delta K_{nonl,e}$ of finite element is not equall to 0, i.e.

$$\Delta K_{nonl,e} \neq 0 \ . \qquad (1.14)$$

If the linear analysis is carried out then the additional finite element is absent and its stiffness matrix is equal to 0, i.e.

$$\Delta K_{nonl,e} = 0\,. \tag{1.15}$$

Schemes of action of the additional finite element are given in Fig. 1.8.

The main function of additional finite element is transformation from the finite element with linear properties to the same finite element but with nonlinear properties (Fig. 1.8a).

From here follows that the inverse process is possible too, i.e. a finite element with nonlinear properties may be transformed into a finite element with linear properties by means of additional finite element (Fig.1.8b).

Besides the third scheme of action of additional finite element exists too (Fig. 1.8c). It is necessary for formation of additional load for the allowance of the nonlinear properties of the considered structure.

Precisely this way opens the possibility for development of the universal method of formation of such loads for nonlinear analysis by finite element method. In analysis of the additional finite element action it ought to know exactly according to what of three enumerated above schemes of action it acts.

a) The transformation from a design diagram consisting of finite elements with linear properties to a design diagram consisting of finite elements with nonlinear properties

Finite element with nonlinear properties	=	Finite element with linear properties	+	Additional finite element for the allowance of the nonlinear properties

b) The transformation from a design diagram consisting of finite elements with nonlinear properties to a design diagram consisting of finite elements with linear properties

Finite element with linear properties	=	Finite element with nonlinear properties	–	Additional finite element for the allowance of the nonlinear properties

c) The formation of additional load for total design diagram consisting of finite elements with nonlinear properties

Additional finite element for the allowance of the nonlinear properties	=	Finite element with nonlinear properties	–	Finite element with linear properties

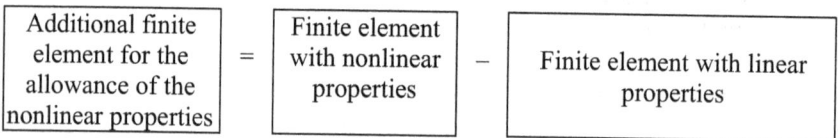

Fig. 1.8. Schemes of action of additional finite elements for the allowance of the nonlinear properties

1.4.2. Formation of additional finite elements

Often in analysis it is necessary to take into account several nonlinear properties. In this case the properties of each additional finite element ought to be formed in accordance with the considered nonlinear properties and then the stiffness matrix of additional finite element $\Delta K_{nonl,e}$ is determined in the following way:

$$\Delta K_{nonl,e} = \sum_{i=1}^{n} \Delta K_{nonl,e,i} , \tag{1.16}$$

where n = number of types of nonlinear properties of the given finite element;

$\Delta K_{nonl,e,i}$ = stiffness matrix of the considered additional finite element which takes into account a manifestation of the i-th nonlinear property.

Stiffness matrix of additional finite element $\Delta K_{nonl,e,i}$ determines the degree of influence of the i-th nonlinear property on behavior of the main finite element.

In view of it the stiffness matrix of additional finite element is determined by the next formula:

$$\Delta K_{nonl,e,i} = K_{nonl,e,i} - K_{nonl,e,i-1} , \tag{1.17}$$

where $K_{nonl,e,i}$ = stiffness matrix of the main finite element with regard for the i-th nonlinear property;

$K_{nonl,e,i-1}$ = stiffness matrix of the main finite element without regard for the i-th nonlinear property.

In general in the presence of the given nonlinear property a stiffness matrix of the additional finite element $\Delta K_{nonl,e,i}$ is not equal to 0, i.e.

$$\Delta K_{nonl,e,i} \neq 0 . \tag{1.18}$$

If the i-th nonlinear property does not influence on behavior of the main finite element then a stiffness matrix of the corresponding finite element $\Delta K_{nonl,e,}$ is equal to 0, i.e.

$$\Delta K_{nonle,i} = 0 . \tag{1.19}$$

Allowance for the first nonlinear property the stiffness matrix of additional finite element $\Delta K_{nonl,e,1}$ may be determined by the next formula:

$$\Delta K_{nonl,e,1} = K_{nonl,e,1} - K_{nonl,e,0} = K_{nonl,e,1} - K_e , \tag{1.20}$$

where K_e = stiffness matrix of the main finite element with linear property.

In case if analysis is carried out without allowance for nonlinear properties of the main finite element the stiffness matrix of the additional finite element $\Delta K_{nonl,e}$ is not taken into account either and it is equal to 0:

$$\Delta K_{nonl,e} = 0 , \tag{1.21}$$

All the other finite elements of this stiffness matrix are equal to 0 too (1.19).

Stiffness matrix of additional finite element $\Delta K_{nonl,e}$ is formed on the basis of nonlinear properties appearing in behavior of the main finite element.

It means that the properties of the additional finite element ought to be determined depending on the number of considered nonlinear properties n, i.e. each additional finite element ought to consist of n component elements each of which takes into account only the i-th nonlinear property. If any nonlinear property is not taken into account then the corresponding to this property additional finite element is absent too. The general scheme for the allowance of the nonlinear properties of the separate finite element is presented in Fig. 1.9. Example of its formation in analysis of reinforced concrete structures depending on their intense exhibited nonlinear properties is given in Fig. 1.10. Six nonlinear properties (n = 6) enumerated in Table In.1 and Fig.1.6 are taken into account in it.

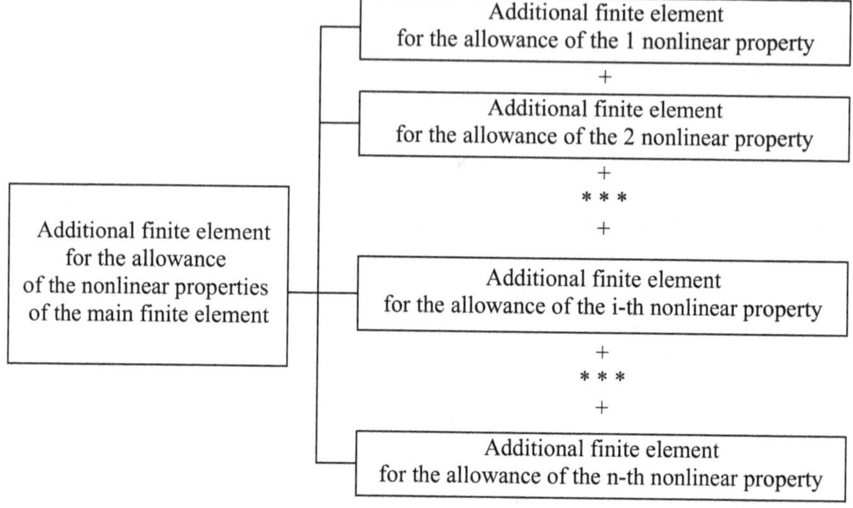

Fig. 1.9. Scheme for the allowance of the nonlinear properties of the separate finite element

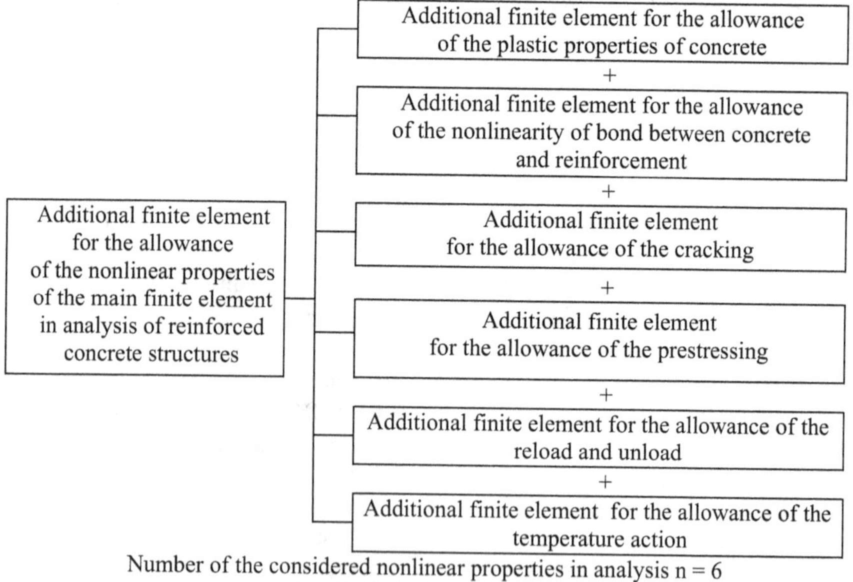

Fig. 1.10. Scheme for the allowance of the nonlinear properties of the separate finite element in analysis of reinforced concrete structures

The stated above approach to formation of additional finite element properties does not introduce any restriction in type, character and number of the considered nonlinear properties of finite elements and finite elements destined for analysis of reinforced concrete structures by finite element method are among them. Besides this approach means that an additional design diagram completely corresponds to the initial design diagram in dimensions and number of component additional finite elements. It takes place because the additional finite elements have shape and size of the main finite element and they change the stiffness characteristics of the main finite element only.

Thus introduction of additional finite elements and additional design diagram does not increase the working time of analysis at the stage of formation of design diagram of the structure. The working time of analysis is increased at the stage of calculation of stiffness matrices of the separate finite elements, as it is necessary to determine stiffness matrices of the main finite elements with linear and nonlinear properties and stiffness matrices of the corresponding additional finite elements.

Besides before the formation of additional finite element properties it is necessary to know the law of deformation of the main finite element with nonlinear properties and the degree of influence each of them on stress-strain state of the given finite element in its behavior from beginning of loading to the moment of its ultimate limit state or the ultimate limit state of structure.

1.4.3 Additional finite element and limit state of the main finite element

The use of additional finite elements for nonlinear analysis of structures creates the problem of determination of properties of such element in case if the main finite element is at ultimate limit state or after it.

Since the behavior of structure at limit state is an extreme type of its nonlinear behavior then the behavior of its separate finite element at limit state is an extreme but nevertheless a special case of its nonlinear behavior.

Therefore schemes of action of additional finite element for the allowance of the limit state of the main finite element presented in Fig. 1.11 in the large ought to correspond to usual action of the additional finite element for the allowance of the nonlinear properties which was given in Fig. 1.8.

After reaching of ultimate limit state the finite element is excluded either partially or completely out of behavior. It means that the additional finite element which is used after limit state ought to work in accordance with two variants.

In the first case the additional finite element excludes the corresponding main finite element partially. Scheme of its behavior is analogous to the presented one in Fig. 1.9.

But the properties of additional finite element for the allowance of the limit state before and after the moment of its reaching have different formation. Up to this moment these properties were determined by all manifested previously nonlinear properties (Fig. 1.12).

After the reaching of limit state they are determined by partial unload due to removal of limit stresses and following manifestation of residual nonlinear properties (Fig. 1.13).

In the second case in total elimination of the main finite element after reaching of ultimate limit state the additional finite element ought to guarantee this process. It means that the main finite element at limit state ought to disappear out of the schemes presented in Fig. 1.11.

a) The transformation from design diagram consisting of finite elements with linear properties to design diagram consisting of finite elements with nonlinear properties

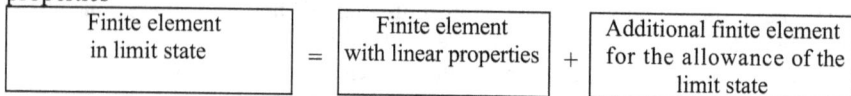

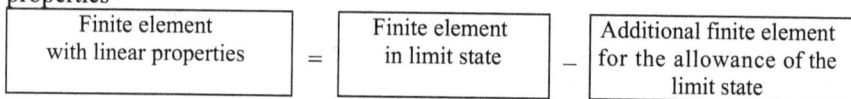

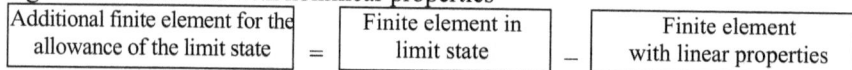

Finite element in limit state	=	Finite element with linear properties	+	Additional finite element for the allowance of the limit state

b) The transformation from design diagram consisting of finite elements with nonlinear properties to design diagram consisting of finite elements with linear properties

Finite element with linear properties	=	Finite element in limit state	–	Additional finite element for the allowance of the limit state

c) The formation of additional loads for total initial design diagram consisting of finite elements with nonlinear properties

Additional finite element for the allowance of the limit state	=	Finite element in limit state	–	Finite element with linear properties

Fig. 1.11. Schemes of action of additional finite elements for the allowance of the limit state

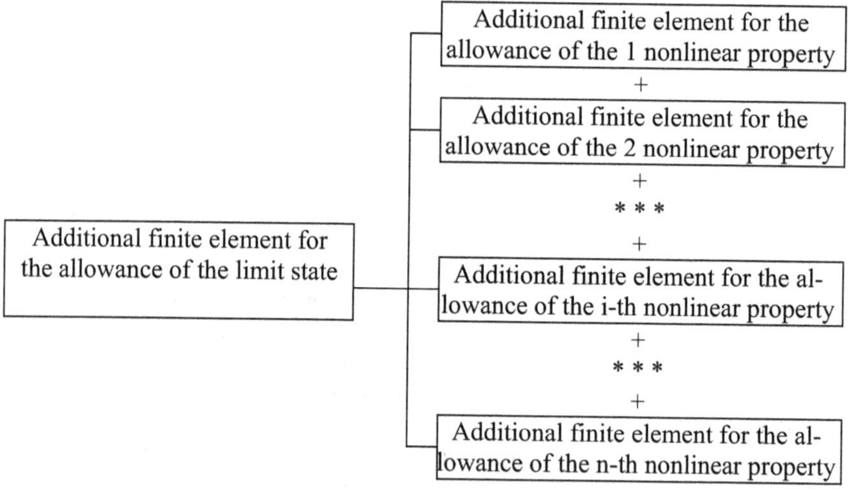

n – number of nonlinear properties considered in the design to moment when the limit state is reached (i changes from 1 to n)

Fig. 1.12. Additional finite element for the allowance of the limit state up to the moment when it is reached

Additional finite element for the allowance of the partial unload
+
Additional finite element for the allowance of the 1 nonlinear property
+
Additional finite element for the allowance of the 2 nonlinear property
+
* * *
+
Additional finite element for the allowance of the i-th nonlinear property
+
* * *
+
Additional finite element for the allowance of the n-th nonlinear property

Additional finite element for the allowance of the limit state

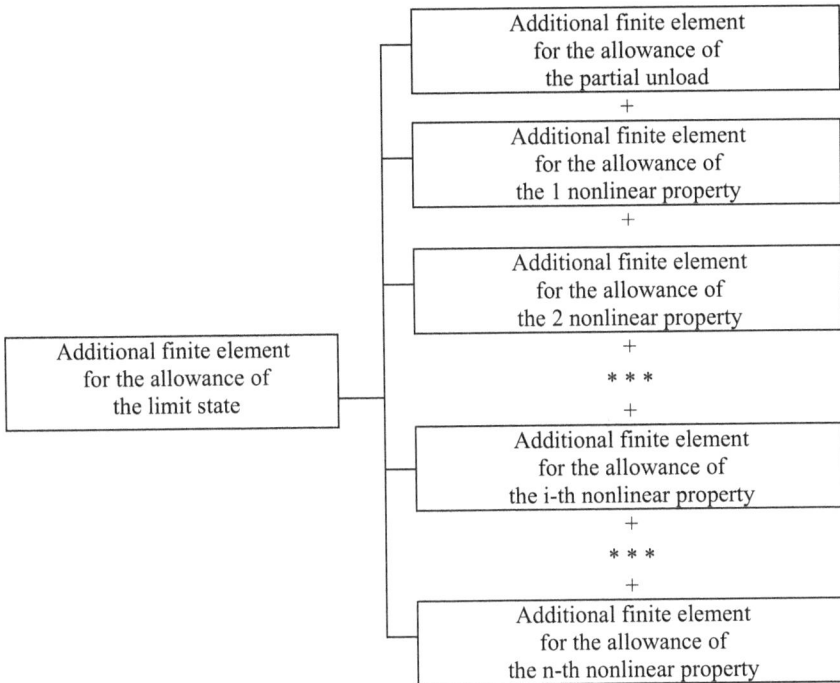

Fig. 1.13. Additional finite element for the allowance of the limit state after the moment when it is reached at partial exclusion of the main finite element

Then in all three cases the properties of additional finite elements are completely determined by the main finite element with linear properties but the action of additional finite element ought to be opposite to the action of the main finite element and correspond to the scheme given in Fig. 1.14.

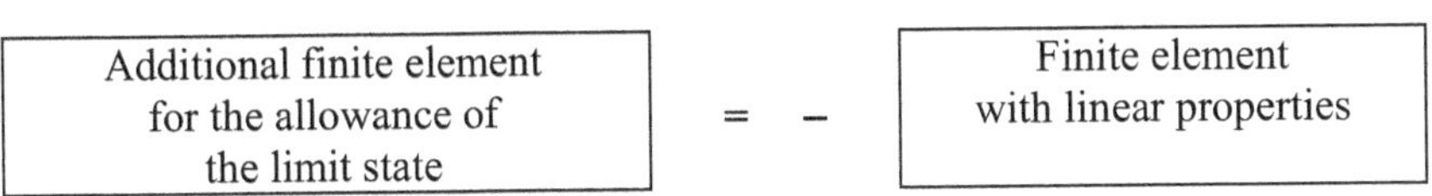

Fig. 1.14. Additional finite element for the allowance of the limit state when it is reached at complete elimination of the main finite element

In previous section (1.2.3) two types of limit state which are characteristic for the triangular concrete deep-beam finite element destined for analysis of reinforced concrete structures at plane stress state by finite element method were considered. It means that the properties of additional finite elements which take into account each state ought to be formed differently depending on the stage (see Fig. 1.7) of behavior of the main finite element. Table 1.2 illustrates this process. It is given below.

Table 1.2

Formation of properties of additional triangular concrete deep-beam finite element for the allowance of the limit state

N	Stage of behavior of the main concrete finite element (see section 1.2.3)	Corresponding additional finite element (AFE)	Account of behavior at limit state		Theoretical sources for development of properties of additional finite elements
			Compression	Tension	
1	Plastic behavior	Additional finite element for the allowance of the plastic properties	+	+	Finite element with plastic properties (see Section In.3.1)
2	Partial unload due to cracking	Additional finite element for the allowance of the residual strains under unload	–	+	See Section 3.2.2.
3	Behavior with crack under reload	Additional finite element for the allowance of the behavior with crack	–	+	Finite element with crack (see Section In. 1.13)
4	Collapse	Additional finite element for the allowance of the collapse	+	+	Fig. 1.13

This table shows the order of formation of the properties of an additional concrete triangular deep-beam finite element on the basis of its limit states. However these states are characteristic for concrete deep-beam finite element of different shape, for example, rectangular one. This is explained by general behavior of concrete at plane stress state. Therefore Table 1.2 shows how the properties of additional concrete rectangular deep-beam finite element are formed too.

1.5. Properties of additional finite elements

It is necessary to know mathematical relationships which characterize the properties of additional finite elements for their application in nonlinear analysis of structures at limit state.

Since the properties of additional finite elements are determined by the properties of a corresponding main finite element then the desired mathematical relationships are determined by analogous relationships of the main finite element [71, 85]:

1) Relationship between node reactions and displacements $\Delta R_e = f(V)$;
2) Formula for determination of stiffness matrices ΔK_e;
3) Formula for determination of node reactions ΔR_e;
4) Relationship between node reactions and stresses $\Delta R_e = f(\sigma)$;

5) Relationship between node reactions and strains $\Delta R_e = f(\Delta\varepsilon)$;
6) Formula for determination of stresses $\Delta\sigma$;
7) Formula for determination of strains $\Delta\varepsilon$;
8) Formula for determination of additional load F_e.

There are two types of additional finite elements as it will be shown later. Not every one of enumerated above relationships is needed for description of each type of finite element but only the relationships characteristic for the given case. In this connection some of the mentioned above relationships are not used for the further analysis.

All enumerated above relationships ought to be determined for each type of using additional finite elements depending on the stage in behavior of the main finite element at reaching the characteristic for the given case of its limit state.

For determination of additional finite elements which are necessary for solving of specific problem the next formula may be used:

$$n_{el} = \sum_{i=1}^{n} n_i \; , \qquad (1.22)$$

where n_{el} = number of additional of additional finite elements; n = number of characteristic limit states of the main finite element; n_i = number of stages in behavior of the main finite element at each the i-th limit state of the total number of its limit states ($i = 1 \div n$).

For example, two types of limit states of strength are characteristic for triangular concrete deep-beam finite element (see Table 1.2), i.e. $n = 2$. At the first of them (compression) the main finite element passes through two states, i.e. $n_1 = 2$. At the second limit state (tension) it passes four states, i.e. $n_2 = 4$. In this case the number of additional finite elements will be equal to $n_{el} = 2 + 4 = 6$.

Such approach to determination of properties of the additional finite elements is defined by requirements for realization of analysis at limit state by finite element method.

1.5.1. Two types of additional finite elements

The appointment of additional finite element is the change of stress-strain state of the main finite element without allowance for nonlinear property to the level of stress-strain state which is appeared in the same finite element with allowance for this property. Such approach opens the opportunity to use of the two ways of change of properties of the main finite element in view of appearance of the definite nonlinear property: change of its initial stress state and change of its initial strain state.

The existence of these two ways is illustrated by example of the first nonlinear property in Fig. In.15, which is presented in Section 3.1. In this case the initial data are the data of behavior of the main finite element with linear properties under the action of outer load P and under the action of outer and additional load in common $(P + F)$.

So if all parameters of linear analysis (line OBC) are known, then it is possible to reach the parameters of nonlinear analysis (point A) by the two ways: by correction of elastic stresses (line OBCA) and by change of elastic strains (line OBA). It means that it is possible to use two variants of change of the main finite element with linear properties too:

1) to change its stress state, i.e. after the reaching of point C corresponding to the nonlinear strains ε_{nonl} and linear stresses σ to change the values of these stresses till the value σ_{nonl} (point A);

2) to change its strain state, i.e. after reaching of point B corresponding to nonlinear stress σ_{nonl} and linear strains ε to change the values of these strains till the value ε_{nonl} (point A).

Since the additional finite element changes the properties of the main finite element then it is recommended to use two types of additional finite elements for realization of each of the two indicated ways:

1) additional finite element of the first type changes a stress state of the main finite element and it does not change its strain state;

2) additional finite element of the second type changes a strain state of the main finite element and it does not change its stress state.

The presence of the two types of additional finite elements is closely connected with requirements to the developed technique of limit state analysis by finite element method. So the first way reflects the singularity of the fifth requirement and the second way reflects the singularity of the fourth requirement (see Section 1.1).

Besides the use of additional finite element of the first type may be realized by method of elastic solutions (additional loads) in the variant of method of additional stresses. This method may be realized in the variant of the method of additional strains (see Section 2.3.1) in use of the additional finite element of the second type.

General approach to the way of formation of additional load with allowance for nonlinear properties for both types of additional finite elements remains the same. Nevertheless there are individual peculiarities which insert simplifications in algorithm of problem solving in comparison with common case.

1.5.2. Additional finite element of the first type

The additional finite element of the first type does not change strain state. It changes only the stress state of the main finite element without account of particular nonlinear property up to the level of stress state of the same element with account of the given nonlinear property.

It is known that stiffness matrix of any finite element connects its node forces and displacements.

Thus for the main finite element at definite stage of its behavior at particular limit state for the allowance of the i-th nonlinear property this relation looks like this:

$$K_{nonl,e,i}\, V = R_{nonl,e,i} \,, \tag{1.23}$$

where $R_{nonl,e,i}$ = node reactions in the finite element for the allowance of the i-th nonlinear property; V = node displacements; $K_{nonl,e,i}$ = stiffness matrix of finite element for the allowance of the first nonlinear property.

Let us express the stiffness matrix $K_{nonl,e,i}$ of the finite element with nonlinear properties using formula (2.17):

$$K_{nonl,e,i} = K_{nonl,e,i1} + \Delta K_{nonl,e,i} \ . \tag{1.24}$$

Node reactions $R_{nonl,i}$ of finite element for the allowance of the i-th nonlinear property may be expressed by reactions $R_{nonl,i-1}$ of the finite element without this i-th nonlinear property:

$$R_{nonl,i} = R_{nonl,i-1} + \Delta R_{nonl,i} \ , \tag{1.25}$$

where $\Delta R_{nonl,i}$ = change of node reactions of finite element due to manifestation of the i-th nonlinear property.

The first component of the right-hand part of this expression presents node reactions in the finite element without allowance for the i-th nonlinear property, i.e.

$$K_{nonl,e,i-1} V = R_{nonl,i-1} \ . \tag{1.26}$$

If we substitute (2.24) and (2.25) in (2.23) with allowance for (2.26) we obtain:

$$\Delta K_{nonl,e,i} V = \Delta R_{nonl,i} \ . \tag{1.27}$$

This formula determines the dependence between node reactions and node displacements in additional finite element for the allowance of the i-th nonlinear property. Its stiffness matrix $\Delta K_{nonl,e,i}$ is determined by formula (1.17) and for the allowance of the first nonlinear property it is determined by formula (1.20).

If the influence of any nonlinear property on behavior of the main finite element is absent then the stiffness matrix of the corresponding additional finite element is equal to 0 (1.19).

If we compare formulae (In.15) and (1.27), we may make a conclusion that by means of node reactions of additional finite elements it is possible to form the vector of additional load for the main finite element, by means of which it is possible to allow for the considered nonlinear property. This vector $F_{e,i}$ is determined by means of the next relationship:

$$F_{e,i} = - \Delta K_{nonl,e,i} V = - \Delta R_{nonl,i} \ . \tag{1.28}$$

Further it follows that we come to a little stop on the case of formation of additional load for collapse of the main finite element after reaching of its ultimate limit state.

In this case let us note the stiffness matrix of additional finite element as ΔK_{lim} and then (see Fig. 1.14):

$$\Delta K_{lim} = - \Delta K \ . \tag{1.29}$$

where K = stiffness matrix of the main finite element with linear properties.

Node reactions of additional finite element ΔR_{lim} are determined by next formula:

$$\Delta R_{lim} = - \Delta K_e V = - R \ , \tag{1.30}$$

where R = node reactions of finite element with linear properties. In this connection the additional load for collapse of finite element is equal to:

$$F_e = K_e V = R \ , \tag{1.31}$$

Actually it means that after collapse the main finite element ceases its existence and does not influence the neighboring elements and additional load guarantees this fact. On the basis of these vectors $F_{e,i}$ the vector of additional load F_i

for the allowance of the i-th nonlinear property of the total structure according its design diagram may be formed. Then on the basis of separate vectors F_i for each considered in analysis nonlinear property the total vector F for all considered nonlinear properties may be formed (see formula In.22).

The absence of strains and the presence of stresses are characteristic for the first type of additional finite element, i.e

$$\Delta\varepsilon_{nonl,i} = 0 , \tag{1.32}$$

$$\Delta\sigma_{nonl,i} \neq 0 . \tag{1.33}$$

It means that if the additional finite element of the first type is used for the allowance of the i-th nonlinear property of the main finite element next formulae are correct for determination of stresses and strains:

$$\sigma_{nonl,i} = \sigma_{nonl,i1} + \Delta\sigma_{nonl,i} , \tag{1.34}$$

$$\varepsilon_{nonl,i} = \varepsilon_{nonl,i-1} . \tag{1.35}$$

These relationships for the allowance of the first nonlinear property look like these ones:

$$\sigma_{nonl,1} = \sigma + \Delta\sigma_{nonl,1} , \tag{1.36}$$

$$\varepsilon_{nonl,1} = \varepsilon . \tag{1.37}$$

It should be noted that general formula (1.28) for calculation of additional load $F_{e,i}$ for the main finite element is correct for both types of additional finite elements. However besides this general approach there are individual cases of determination of vector of node reactions $\Delta R_{nonl,i}$ and consequently of additional load $F_{e,i}$ for both types of additional finite elements.

These individual cases are based on the relationships between node reactions with stresses and strains. For additional finite element of the first type this individual case is based on relationship between node reactions and stresses.

So for the main finite element with linear properties the relationship between node reactions R and stresses σ looks like:

$$R = C\sigma , \tag{1.38}$$

where C = matrix connecting elastic node reactions and stresses.

The analogous formula for the main finite element with linear properties looks like:

$$R_{nonl} = C\,\sigma_{nonl} . \tag{1.39}$$

Using the relationships (1.38) and (1.39) let us write down the formulae for determinations of node reactions $\Delta R_{nonl,i}$ in additional finite element of the first type for the allowance of the first nonlinear property:

$$\Delta R_{nonl,1} = C\,\Delta\sigma_{nonl,1} , \tag{1.40}$$

and for the allowance of the i-th nonlinear property:

$$\Delta R_{nonl,i} = C\,\Delta\sigma_{nonl,i} . \tag{1.41}$$

Substituting the expression (1.41) in (1.28) we obtain the formula for determination of the vector of additional load in general case:

$$F_{e,i} = -\,C\,\Delta\sigma_{nonl,i} . \tag{1.42}$$

In elimination of the main finite element after reaching of its ultimate limit state the formula looks like:

$$F_{e,i} = C\,\sigma\,, \qquad (1.43)$$

where σ = stresses of finite element with linear properties.

The additional load of the triangular deep-beam finite element is calculated in this way (see Section In 1.3):

$$F_{e,i} = -\,tS(A^{-1})^T B^T\,\Delta\sigma_{nonl,i}\,, \qquad (1.44)$$

and

$$C = \;tS(A^{-1})^T B^T\;. \qquad (1.45)$$

This simplified variant of formation of the additional load using stresses $\Delta\sigma_{nonl,i}$ for the allowance of the plastic properties of concrete was developed in research [50] and used in investigations [25, 94]. The main characteristics of additional finite element of the first type for the allowance of the nonlinear property are presented briefly in Section 1.5.4. The analogous relationships for the main finite element with linear properties at the first nonlinear property and at collapse are given there too.

1.5.3. Additional finite element of the second type

Additional finite element of the second type does not change stress state. It changes only strain state of the main finite element without allowance of the nonlinear property up to the level of strain state for the allowance of the given nonlinear property.

The dependence (1.27) connecting node reactions and node displacements as it reflects the main property of any additional finite element is correct for additional finite element of the second type too. Stiffness matrix of this element may be determined by formula (1.17) and its additional load for the allowance of the nonlinear property may be determined by formula (1.28). These expressions are correct for additional finite element of any type too. Relating to other formulae describing the properties of additional finite element of the second type that it is necessary to use another approach for their obtaining based on the peculiarities of this element.

The absence of stresses and presence of strains is a defining feature of the additional finite element of the second type, i.e.

$$\Delta\varepsilon_{nonl,i}. \neq 0\,, \qquad (1.46)$$

$$\Delta\sigma_{nonl,\,i} = 0\,. \qquad (1.47)$$

It means that if the additional finite element of the second type is used for the allowance of the i-th nonlinear property of the main finite element the next formulae for determination of stresses and strains are correct:

$$\sigma_{nonl,i} = \sigma_{nonl,i}\,(\varepsilon_{nonl,i-1})\,, \qquad (1.48)$$

$$\varepsilon_{nonl,i} = \varepsilon_{nonl,i-1} + \Delta\varepsilon_{nonl,i}\,. \qquad (1.49)$$

They are correct for the first nonlinear property:

$$\sigma_{nonl,1} = \sigma(\varepsilon), \tag{1.50}$$

$$\varepsilon_{nonl,1} = \varepsilon + \Delta\varepsilon_{nonl,1}. \tag{1.51}$$

Special case of determination of additional load for additional finite element of the second type is based on the relationship of node reactions and strains of the main finite element.

For the main finite element with linear properties this relationship has next form:

$$R = G\,\varepsilon\ , \tag{1.52}$$

where G = matrix connecting elastic node reactions and strains.

For the main finite element with nonlinear properties this relationship looks like:

$$R_{nonl} = G\,\varepsilon_{nonl}\ . \tag{1.53}$$

Taking into consideration formulae (1.52) and (1.53) the formula for determination of node reaction in additional finite element of the second type for the allowance of the first nonlinear property will be:

$$\Delta R_{nonl,1} = G\,\Delta\,\varepsilon_{nonl,1}\ . \tag{1.54}$$

This formula for the allowance of the i-th nonlinear property looks like:

$$\Delta R_{nonl,i} = G\,\Delta\varepsilon_{nonl,i}\ . \tag{1.55}$$

Using the formulae (1.55) and (1.28) we may obtain the formula for determination of the additional load at any stage:

$$F_{e,i} = -\,G\,\Delta\varepsilon_{nonl,i}\ . \tag{1.56}$$

In case of collapse of the main finite element when its ultimate limit state is reached this formula will be of next form:

$$F_e = G\varepsilon\ , \tag{1.57}$$

where ε = strains of the finite element with linear properties. For triangular deep-beam finite element the additional load (see Section 1.1.3) is determined by next formulae:

$$F_{e,i} = -\,tS(A^{-1})^T B^T\,D\Delta\varepsilon_{nonl,i} \tag{1.58}$$

and

$$G = tS(A^{-1})^T B^T D = CD\ . \tag{1.59}$$

This variant of formation of additional load $F_{e,i}$ using the strains $\Delta\varepsilon_{nonl,i}$ of additional finite element is suitable to take into account residual strains under unloading that will be shown later.

Formulae (1.54) and (1.55) may be used only for additional finite element of the second type as for additional finite element of the first type it is accepted that its strains $\Delta\varepsilon_{nonl,i} = 0$ (1.32). The algorithm of the problem solving may be shorted by these formulae.

Briefly the main characteristics of the additional finite elements of the second type for the allowance of the first nonlinear property are given in the next section. The analogous formulae for the main finite element with linear properties in the presence of the nonlinear property and after collapse of the main finite element after reaching of its ultimate limit state are given there too.

1.5.4. Brief information of the main characteristics of additional finite elements

Brief information of the main relationships describing the properties of additional finite elements of both types with allowance for the first and any other nonlinear property as well as collapse after reaching of ultimate limit state are given in Tables 1.3, 1.4, 1.5.

Table 1.3

The main characteristics of additional finite elements for the allowance of the first nonlinear property

№	Type of characteristic	Finite element with linear properties	Finite element with the first nonlinear property	Additional finite element for the allowance of the first nonlinear property	
				Type 1	Type 2
1	Relationship between node reactions and displacements	$R = K_e V$	$R_{nonl,1} = K_{nonl,e,1} V$	$\Delta R_{nonl,1} = \Delta K_{nonl,e,1} V$	
2	Stiffness matrices	K_e	$K_{nonl,e,1}$	$\Delta K_{nonl,e,1} = K_{nonl,e,1} - K_e$	
3	Node reactions	R	$R_{nonl,1}$	$\Delta R_{nonl,1} = R_{nonl,1} - R$	
4	Relationship between node reactions and stresses	$R = C\sigma$	$R_{nonl,1} = C \sigma_{nonl,1}$	$\Delta R_{nonl,1} = C \Delta\sigma_{nonl,1}$	–
5	Relationship between node reactions and strains	$R = G\varepsilon$	$R_{nonl} = G\varepsilon_{nonl}$	–	$\Delta R_{nonl,1} = G\Delta\varepsilon_{nonl,1}$
6	Stresses	σ	$\sigma_{nonl,1}$	$\Delta\sigma_{nonl,1} = \sigma_{nonl,1} - \sigma$	$\Delta\sigma_{nonl,1} = 0$
7	Strains	ε	$\varepsilon_{nonl,1}$	$\Delta\varepsilon_{nonl,1} = 0$	$\Delta\varepsilon_{nonl,1} = \varepsilon_{nonl,1} - \varepsilon$
8	Additional load	$F_{e,1} = -\Delta R_{nonl,1}$	–	–	–

Table 1.4

The main characteristics of additional finite elements for the allowance of the *i*-th nonlinear property

№	Type of characteristic	Finite element without the *i*-th nonlinear property	Finite element with the *i*-th nonlinear property	Additional finite element for the allowance of the *i*-th nonlinear property	
				Type 1	Type 2
1	Relationship between node reactions and displacements	$R_{nonl,i-1} = K_{nonl,e,i-1}V$	$R_{nonl,i} = K_{nonl,e,i}V$	$\Delta R_{nonl,i} = \Delta K_{nonl,e,i}V$	
2	Stiffness matrices	$K_{nonl,e,i-1}$	$K_{nonl,e,i}$	$\Delta K_{nonl,e,i} = K_{nonl,e,i} - K_{nonl,e,i-1}$	
3	Node reactions	$R_{nonl,i-1}$	$R_{nonl,i}$	$\Delta R_{nonl,i} = R_{nonl,i} - R_{nonl,i-1}$	
4	Relationship between node reactions and stresses	$R_{nonl,i-1} = C\sigma_{nonl,i-1}$	$R_{nonl,i} = C\,\sigma_{nonl,i}$	$\Delta R_{nonl,i} = C\,\Delta\sigma_{nonl,i}$	–
5	Relationship between node reactions and strains	$R_{nonl,i-1} = G\varepsilon_{nonl,i-1}$	$R_{nonl,i} = G\varepsilon_{nonl,i}$	–	$\Delta R_{nonl,i} = G\Delta\varepsilon_{nonl,i}$
6	Stresses	$\sigma_{nonl,i-1}$	$\sigma_{nonl,i}$	$\Delta\sigma_{nonl,i} = \sigma_{nonl,i} - \sigma_{nonl,i-1}$	$\Delta\sigma_{nonl,i} = 0$
7	Strains	$\varepsilon_{nonl,i-1}$	$\varepsilon_{nonl,i}$	$\Delta\varepsilon_{nonl,i} = 0$	$\Delta\varepsilon_{nonl,i} = \varepsilon_{nonl,i} - \varepsilon_{nonl,i-1}$
8	Additional load	$F_{e,i} = -\Delta R_{nonl,i}$	–	–	–

Table 1.5

The main characteristics of additional finite elements for the allowance of collapse of the main finite element after the reaching of ultimate limit state

№	Type of characteristic	Finite element with linear properties	Finite element after the reaching of ultimate limit state	Additional finite elements for the allowance of collapse of the main finite element	
				Type 1	Type 2
1	Relationship between node reactions and displacements	$R=K_eV$	0	$\Delta R_{lim} = \Delta K_{lim}V$	
2	Stiffness matrices	K_e	0	$\Delta K_{lim} = -K_e$	
3	Node reactions	R	0	$\Delta R_{lim} = -R$	
4	Relationship between node reactions and stresses	$R = C\sigma$	0	$\Delta R_{lim} = -C\sigma$	–
5	Relationship between node reactions and strains	$R = G\varepsilon$	0	–	$\Delta R_{lim} = -G\varepsilon$
6	Stresses	σ	0	$\Delta\sigma_{lim} = -\sigma$	$\Delta\sigma_{lim} = 0$
7	Strains	ε	0	$\Delta\varepsilon_{lim} = 0$	$\Delta\varepsilon_{lim} = -\varepsilon$
8	Additional load	$F_e = R$	–	–	–

Table 1.6

The main characteristics of additional triangular finite element of deep-beam for the allowance of collapse of the main finite element after the reaching of ultimate limit state

№	Type of characteristic	Finite element with linear properties	Finite element after the reaching of ultimate limit state	Additional finite element for the allowance of the collapse of the main finite element	
				Type 1	Type 2
1	Relationship between node reactions and displacements	$R = K_e V$	0	$\Delta R_{lim} = -K_e V$	
2	Stiffness matrices	$K_e = tS(A^{-1})^T B^T DB A^{-1}$	0	$\Delta K_{lim} = -K_e = -tS(A^{-1})^T B^T DB A^{-1}$	
3	Node reactions	R_e	0	$\Delta R_{lim} = -R$	
4	Relationship between node reactions and stresses	$R = CtS(A^{-1})^T B^T D\sigma$	0	$\Delta R_{lim} = -tS(A^{-1})^T B^T \sigma$	–
5	Relationship between node reactions and strains	$R = GtS(A^{-1})^T B^T D\varepsilon$	0	–	$\Delta R_{lim} = -tS(A^{-1})^T B^T \varepsilon$
6	Stresses	$\sigma = DB A^{-1} V$	0	$\Delta\sigma_{lim} = -\sigma = -DBA^{-1} V$	$\Delta\sigma_{lim} = 0$
7	Strains	$\varepsilon = B A^{-1} V$	0	$\Delta\varepsilon_{lim} = 0$	$\Delta\varepsilon_{lim} = -\varepsilon = -B A^{-1} V$
8	Additional load	$F_{e,i} = R$	–	–	–

Choice of the type of additional finite element depends on the next factors: type of the solving nonlinear problem, type of the used finite element, degree of limit state reached by the main finite element, peculiarities of influence of the specific nonlinear property on behavior of the main finite element and mathematical formulae describing this nonlinear property with law of deformation of finite element material among them.

Description of the properties of triangular deep-beam additional finite element for collapse of the main finite element after reaching of ultimate limit state is given in Table 1.6.

Chapter 2. INFLUENCE OF ADDITIONAL FINITE ELEMENTS ON GENERAL ORDER OF NONLINEAR PROBLEM SOLVING BY FINITE ELEMENT METHOD

Theoretical basis of additional finite element method destined for solving of nonlinear problem of limit state analysis with use of properties of the additional design diagram of structure formed from the additional finite elements was developed in previous chapter.

This method as any other influences the algorithms and programs destined for its computer realization.

It requires the presence of special blocks for analysis of behavior of each finite element depending on a stage of achieved limit state, determination of additional finite element properties and formation of a vector of additional loads if this vector is used in nonlinear analysis of reinforced concrete structures.

Limit state method on the basis of use of properties of the main finite elements and additional ones although lays definite peculiarities on the general order of nonlinear analysis of structures by finite element method but does not change it as a whole and consequently the order of program construction but for realization of this analysis.

This chapter is devoted to these problems.

2.1. General algorithms for nonlinear analysis of structures at limit state by additional finite elements

The enlarged design diagram for analysis of structures by finite element method with allowance for linear properties of these structures was given in section In.1.1 in Fig. In.1.

Two ways which take into account nonlinear properties by change of stiffness matrix of structure (Section In.2.2 and Fig. In.6) and by impose of additional load (Section In.2.3 and Fig. In.8) were considered in section In.2.

The realization of analysis by additional finite element method is possible by change of stiffness matrix (the first group of ways for the allowance of the nonlinear properties) and by imposes of additional load (the second group). Design diagrams of analysis of structures by each of these variants are given in Fig. 2.1 and Fig. 2.2.

In both cases the general sequence is solving problem by finite element method remains invariable (Section In.1.1), but the use of additional finite elements influences some steps of the problem solving. At the same time some steps are identical in both variants (№ 1, 6 and 7).

Let us consider in detail the sequence of problem solving for analysis of influence on its solution of additional finite element method according to the design diagram shown in Fig. 2.1 (in parenthesis are the numbers of stages of problem solving in accordance with design diagram given in Fig. 2.2):

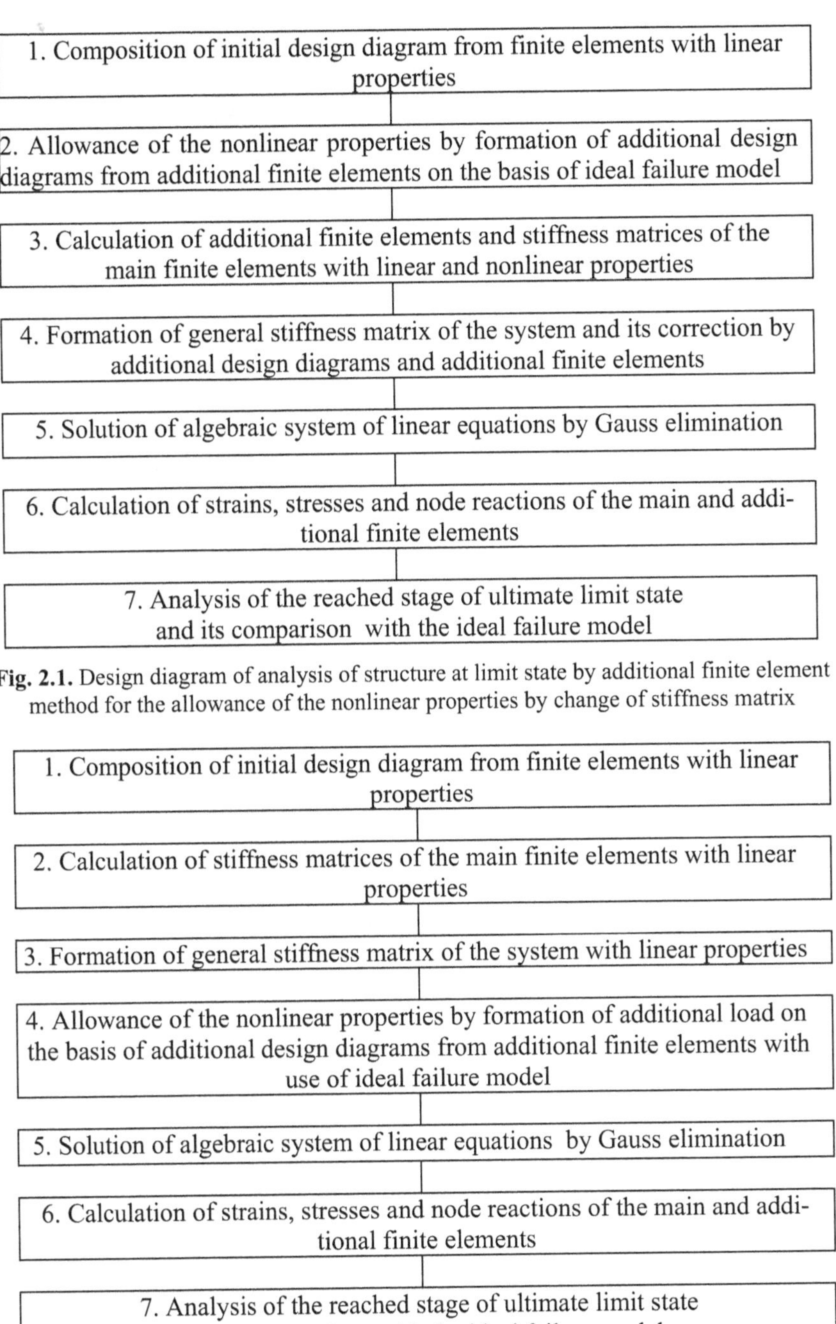
1. Composition of initial design diagram from finite elements with linear properties
2. Allowance of the nonlinear properties by formation of additional design diagrams from additional finite elements on the basis of ideal failure model
3. Calculation of additional finite elements and stiffness matrices of the main finite elements with linear and nonlinear properties
4. Formation of general stiffness matrix of the system and its correction by additional design diagrams and additional finite elements
5. Solution of algebraic system of linear equations by Gauss elimination
6. Calculation of strains, stresses and node reactions of the main and additional finite elements
7. Analysis of the reached stage of ultimate limit state and its comparison with the ideal failure model

Fig. 2.1. Design diagram of analysis of structure at limit state by additional finite element method for the allowance of the nonlinear properties by change of stiffness matrix

1. Composition of initial design diagram from finite elements with linear properties
2. Calculation of stiffness matrices of the main finite elements with linear properties
3. Formation of general stiffness matrix of the system with linear properties
4. Allowance of the nonlinear properties by formation of additional load on the basis of additional design diagrams from additional finite elements with use of ideal failure model
5. Solution of algebraic system of linear equations by Gauss elimination
6. Calculation of strains, stresses and node reactions of the main and additional finite elements
7. Analysis of the reached stage of ultimate limit state and its comparison with the ideal failure model

Fig. 2.2. Design diagram of analysis of structure at limit state by additional finite element method for the allowance of the nonlinear properties by use of additional load

1. Composition of initial design diagram from finite elements with linear properties. Design diagram of structure is constructed by usual way from the main linear elements. Since each main finite element has a corresponding additional nonlinear finite element which has the same size and shape as the main finite element then the assembly of additional finite elements is completely analogous to the assembly of the main ones. It means that it is not necessary to construct the auxiliary design diagram from additional nonlinear finite elements aside from the main design diagram. The auxiliary design diagram at this stage is a copy in geometry of the initial design diagram. This step of problem solving coincides in both cases.

2(4). Allowance of the nonlinear properties by formation of the additional design diagrams from additional finite elements on the basis of ideal failure model. This operation is not fulfilled at the first iteration of the first step of loading but it is fulfilled at the rest steps.

This operation for the allowance of the nonlinear properties by change of the stiffness matrix of structure is fulfilled at once after the preparation of design diagram from the additional finite elements in accordance with the ideal failure model.

This operation for the allowance of the nonlinear properties by formation of the additional load F_{nonl} is fulfilled after the formation of general stiffness matrix with linear properties, i.e. it is the fourth operation and it includes the formation of additional load on the basis of additional design diagram and node reactions of the additional finite elements in accordance with accepted ideal failure model of structure.

3(2). Calculation of stiffness matrices of the separate finite elements.

This stage is the third one in use of the first way for the allowance of the nonlinear properties or the second one in use of the second way for the allowance of the properties. In the first case the stiffness matrices of the main linear finite elements K_e, nonlinear finite elements $K_{nonl,e,i}$ and additional nonlinear finite elements $\Delta K_{nonl,e,i}$ are calculated at all iterations of this step. In the second case this operation is fulfilled at the first iteration of the first step of analysis only. Stiffness matrices of the additional nonlinear finite elements $\Delta K_{nonl,e,i}$ are determined only at the rest iterations of the first step and all iterations of the consequent steps in use of the additional load.

4(3). Formation of general stiffness matrix of the system.

This stage is fulfilled as the fourth step in change of the general stiffness matrix and it is the third step in use of the additional load. In the first case the general stiffness matrix of structure is formed and it is corrected by the stiffness matrices of additional finite elements. In the second case when additional load is used at this step of problem solving two stiffness matrices of the structure are formed: stiffness matrix from the main linear finite elements K and stiffness matrix from the additional finite elements ΔK_{nonl}, but these operations are fulfilled only at the first iteration of the first step. The additional nonlinear stiffness matrix of structure ΔK_{nonl} is formed at the rest iterations of the first step and all iterations of the rest steps only. The stiffness matrix from additional finite elements may be not formed at all if the algorithm of formation of the vector of additional load F_{nonl} on the basis of properties of the additional finite elements is used. In addition in both cases the values of node displacements are taken from the previous iteration.

5. Solution of algebraic system of linear equations by Gauss elimination.

This operation is usually fulfilled in two stages: direct execution and inverse one. The direct execution is connected with the transformation of system of equations; the inverse execution is connected with calculation of unknowns. If the additional load F_{nonl} is used in nonlinear problem solving by finite element method three stages should be distinguished:

1) operations of direct execution of Gauss method for calculation of inverse linear stiffness matrix of structure K^{-1};

2) operations of direct execution of Gauss method for transformation of the column of constant terms P or $P + F_{nonl}$;

3) operations of inverse execution of Gauss method for calculation of unknown node displacements V.

These three operations for the allowance of the nonlinear properties by the first way, i.e. by change of general stiffness matrix of structure are fulfilled at all iterations of all steps of loading. In realization of allowance for nonlinear properties by the second way, i.e. by impose of additional loads all the three operations are fulfilled at the first iteration of the first step of load only. At the rest iterations of this step and all iterations of the rest steps it is sufficiently to fulfill the second operation and the third one only. General design diagram of the system of equations was given in section In.2.3.2 in Fig.In.9.

6. Calculations of strains, stresses and node reactions.

Determination of the values of strains (ε, ε_{nonl}, $\Delta\varepsilon_{nonl}$), stresses (σ, σ_{nonl}, $\Delta\sigma_{nonl}$) and node reactions (R, R_{nonl}, ΔR_{nonl}) is fulfilled in the main linear and nonlinear finite elements, and in the additional finite elements at each step of loading after the end of iterative process. This operation is the same in both cases for the allowance of the nonlinear properties.

7. Analysis of stress-strain state.

This analysis is fulfilled at each step of loading after the end of determination of all necessary values.

Schemes of strains, curves of stresses, domains of stresses, schemes of cracking may be constructed at this stage. Processing of other necessary parameters may be fulfilled there too. Comparison of the obtained stress-strain state with the ultimate limit state corresponding to ideal failure model may be done at this stage too. This comparison is carried out by matching of the ideal failure model with the design diagram of structure at this stage which is a result of action of the additional design diagram on the initial design diagram from linear elements. Operations of this step are the same in both variants of allowance for nonlinear properties.

The conducted analysis allows to estimate the influence of additional finite element method on the order of nonlinear problem solving in analysis of structure by finite element method and leads to conclusion that it is appropriate to realize analysis with impose of additional load constructed on the basis of node reactions of additional finite elements for the allowance of the nonlinear properties. This way leads to the rational construction of iterative process in the problem solving.

2.2. Construction of programs for nonlinear analysis of reinforced concrete structures at limit state by additional loads

It was shown in the previous section that algorithm created on the basis of additional load determined by node reactions of additional finite elements is more suitable to use in construction of programs for nonlinear analysis at limit state. Further it is necessary to consider the enlarged design diagram of program construction for realization of this algorithm.

The design diagram for analysis of structures and therefore for program construction of structural analysis by finite element method for the allowance of the nonlinear properties by application of additional load F_{nonl} was presented in section In.2.3.2 in Fig. In.8. This design diagram includes sequential fulfillment of the main calculating operations of finite element method.

In realization of the mentioned above algorithm this sequence of operations remains invariable as a whole but some changes connected with the use of additional finite elements and the accepted ideal failure model should be inserted. The enlarged design diagram of such program construction is presented in Fig. 2.3. This design diagram with some peculiarities repeats the design diagram which was given in Fig. 2.2 and described in previous section.

The main difference is the appearance in the beginning of analysis of the block "Input data and control of initial information including information of ideal failure model" after the block "Composition of initial design diagram from finite elements with linear properties" and at the end of analysis of block "Output of results of analysis" in front of the block "Analysis of results by determination of the reached degree of ultimate limit state and its comparison with ultimate limit state of ideal failure model".

Besides the block "Calculation of stiffness matrices of the main finite elements with linear properties" is absent as usually in majority of programs this block is united with the block "Formation of general stiffness matrix of system with linear properties".

Let us mark the block "Input data and control of initial information including information of ideal failure model". At this stage the input data for linear and nonlinear analysis including information of ideal failure model and its control with a view to search for the possible errors takes place. Each program has this block.

Block "Output of results of analysis" is a necessary one for each program. In problem solving it includes two types of output information: the tabular type (for output of values of node displacements V, stresses ε_{nonl}, strains σ_{nonl}, node reactions R_{nonl} and other numerical values) and the graphical type (for constructions of schemes of strains, curves of stresses and domains of stresses, schemes of cracking, schemes of disposition of finite elements at ultimate limit state and other graphic information).

Information obtained at this step of problem solving helps to control results with a view to determination of stress-strain state of the structure and its comparison with ultimate limit state defined by accepted ideal failure model of the structure.

The fulfilled analysis allows to estimate the character of influence of additional loads on the order of program construction for realization of nonlinear analysis by finite element method.

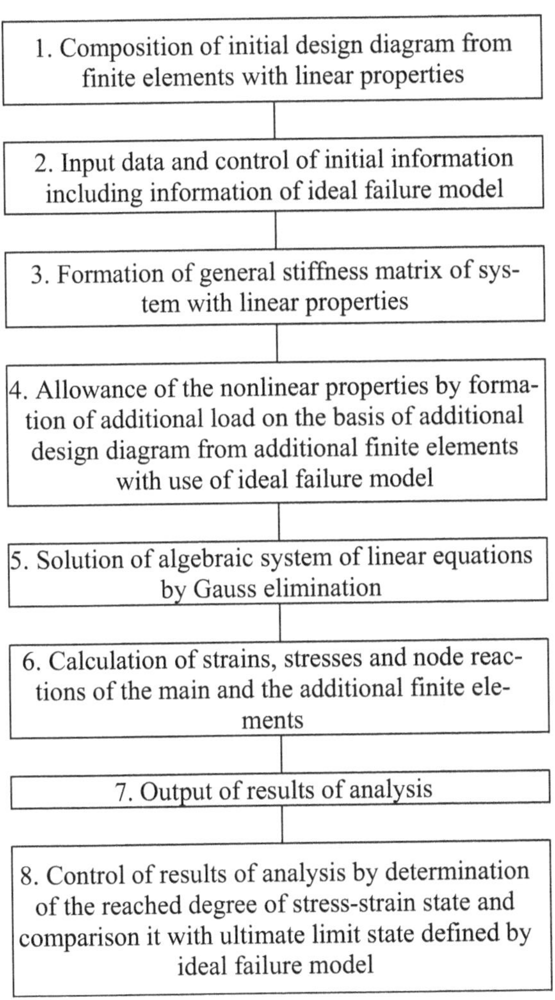
1. Composition of initial design diagram from finite elements with linear properties
2. Input data and control of initial information including information of ideal failure model
3. Formation of general stiffness matrix of system with linear properties
4. Allowance of the nonlinear properties by formation of additional load on the basis of additional design diagram from additional finite elements with use of ideal failure model
5. Solution of algebraic system of linear equations by Gauss elimination
6. Calculation of strains, stresses and node reactions of the main and the additional finite elements
7. Output of results of analysis
8. Control of results of analysis by determination of the reached degree of stress-strain state and comparison it with ultimate limit state defined by ideal failure model

Fig.2.3. Design diagram of construction of program realizing analysis of structures at limit state by additional finite element method for the allowance of the nonlinear properties by additional load

2.3. Singularities of iteration process of computer realization of problem solving

The construction of iterative process depending on the stage of problem solving is one of the main problems connected with computer realization of non-linear analysis of structures at limit state with use of additional loads. Two aspects of the problem may be distinguished: the character of iterative process construction depending on the stage of problem solving and the test of process end.

Table 2.1 presents the singularities of iterative process construction depending on the stage of problem solving, i.e. it helps to characterize the first aspect of the considered problem.

Table 2.1

Singularities of iterative process in construction of programs realizing an analysis of the structures at limit state for the allowance of the nonlinear properties by additional loads F_{nonl}

№	Contents of stage		Place of fulfillment				Note
	Description	Composition of operations	Beginning of analysis	The 1 iteration of the1step	Other iterations	End of step	
1	2	3	4	5	6	7	8
1	Composition of initial design diagram from finite elements with linear properties	Design diagram from linear finite elements					DD is formed once
		Design diagram from additional finite elements	+				It is a geometric copy of DD from linear FEs
2	Input data and control of initial information including information of ideal failure model	Input data for linear analysis	+				It is fulfilled once
		Control of data for linear analysis	+				
		Input data for nonlinear analysis	+				
		Control of data for nonlinear analysis	+				
		Input data of ideal failure model	+				
		Control of data of ideal failure model	+				
3	Formation of general stiffness matrix of system with linear properties	Stiffness matrices K from linear finite elements		+			It is formed once

1	2	3	4	5	6	7	8
4	Allowance for nonlinear properties by formation of additional load on the basis of additional design diagrams and additional finite elements with use of ideal failure model	Formation of the vector of additional loads F_{nonl} on the basis of additional design diagrams from additional finite elements with use of ideal failure model			+		The vector is calculated repeatedly
5	Solution of algebraic system of linear equations by Gauss elimination	Direct execution of calculation of inverse linear stiffness matrix of structure K^{-1}		+			It is fulfilled once
		Direct execution of transformation of column of constant terms $P + F_{nonl}$		+	+		It is fulfilled repeatedly
		Inverse execution of calculation of unknown displacements V		+	+		It is fulfilled repeatedly
6	Calculation of strains, stresses, and node reactions of the main finite elements and the additional ones and other parameters	Strains ε_{nonl}, $\Delta\varepsilon_{nonl}$				+	It is calculated at the end of each step
7	Output of results of analysis	Displacements V				+	It is in the form of tables at the end of each step
		Strains ε_{nonl}				+	
		Stresses σ_{nonl}				+	
		Node reactions R_{nonl}				+	
		Other parameters of limit state				+	

End of Table 2.1

1	2	3	4	5	6	7	8
		Schemes of deformation				+	It is in the form of graphics at the end of each step
		Curves and domains of stresses				+	
		Schemes of cracking				+	
		Schemes of disposition of finite elements at ultimate limit state				+	
		Treatment of other parameters				+	
8	Control of results of analysis by determination of the reached degree of stress-strain state and its comparison with ideal failure model	Schemes of deformation				+	It is fulfilled at the end of each step
		Curves of stresses				+	
		Domains of stresses				+	
		Schemes of cracking				+	
		Comparison with ideal failure model				+	
		Schemes of finite elements at limit state				+	
		Comparison with ideal failure model of finite element				+	
		Treatment of other parameters				+	
		Data of achieved stage of stress-strain state				+	

In convergent iterative process the reaching of prescribed precision of calculation of node displacements V is the test of its end. In no convergent iterative process a sufficient number of fulfilled iterations may be accepted as its end criterion at reaching which the problem solving is broken. Use of additional load for the allowance of the nonlinear properties exhibited by the structure as its ultimate limit state is reaching influences the iterative process construction but this influence has not the essential character and does not change the main sequence of problem solving.

At the same time the method of additional load itself converges badly under great nonlinear strains (see Section In.2.3.1). Thus according to preliminary estimation the difference between linear solving and nonlinear one ought to be less then 35 – 45 %. This problem requires a further detail study.

Chapter 3. ADDITIONAL FINITE ELEMENTS AND ADDITIONAL LOADS FOR NONLINEAR ANALYSIS OF REINFORCED CONCRETE STRUCTURES

Analysis of reinforced concrete structures by finite element method is connected with taking account of specific nonlinear properties the main of which are next: plasticity of concrete, bond between concrete and reinforcement, cracking, unload and reload, action of temperature and prestressing.

The presence of such different by nature nonlinear properties manifested by structures up to the moment of reaching of their ultimate limit state requires the creation of flexible algorithm with allowance for them. This chapter is devoted to this problem.

3.1. The main types of additional loads for nonlinear analysis of reinforced concrete structures

The properties of additional design diagrams and additional finite elements are used in development of the algorithm of formation of additional loads for nonlinear analysis of reinforced concrete structures at limit state.

3.1.1. General order of formation of additional loads on the basis of additional design diagrams

The description of additional design diagram of structure which allows to form the additional loads for nonlinear analysis (see In. 2.3.1 and In. 3.1) using the method of elastic solving was given in section 1.3.

For formation of additional load with allowance for any nonlinearity next sequence of the main operations is offered:

1. Representation of the main decision system of linear equations in the form $K_{nonl}\ V = P$ (see In. 2 and 1.1).

2. Consideration of general stiffness matrix of structure with nonlinear properties K_{nonl} as the sum of stiffness matrix K from stiffness matrices of finite elements with linear properties and additional stiffness matrix ΔK_{nonl} from stiffness matrices of additional finite elements: $K_{nonl} = K + \Delta\ K_{nonl}$ (see In. 13 and 1.2).

3. Representation of the main decision equation in the form: $(K + \Delta\ K_{nonl}) V = P$.

4. Opening of parentheses and carry of the second term in the right hand part of equation: $K\ V = P - \Delta\ K_{nonl}\ V$ (see In. 14 and 1.40).

5. Determination of the vector of additional load $F_{nonl} = -\ \Delta\ K_{nonl}\ V$ (see In. 15 and 1.5).

Enumeration of the main nonlinear properties of reinforced concrete structures accompanied by notation of the corresponding additional load was given in Table In.1. The sequence of formation of each additional load on the basis of additional design diagrams of structure is given in Table 3.2. The used system of notations is presented in Table 3.1.

Algorithm of formation of additional loads for the allowance of the main nonlinear properties of reinforced concrete structures is based on the use of tools from additional finite elements taking into account of each separate nonlinear property.

It means that all necessary characteristics of these elements ought to be prescribed in advance.

Table 3.1

Values necessary for formation of additional load on the basis of additional design diagrams

№	General title of value	Nonlinear stiffness matrix of structure	Additional stiffness matrix of structure	Additional load
1	2	3	4	5
1	General notation in nonlinearity	K_{nonl}	ΔK_{nonl}	F_{nonl}
2	Plastic properties of concrete	K_{pl}	ΔK_{pl}	F_{pl}
3	Nonlinearity of bond between concrete and reinforcement	K_{bond}	ΔK_{bond}	F_{bond}
4	Cracking	K_{crc}	ΔK_{crc}	F_{crc}
5	Reload and unload	K_{rep}	ΔK_{rep}	F_{rep}
6	Action of temperature	K_t	ΔK_t	F_t
7	Prestressing	K_{pr}	ΔK_{pr}	F_{pr}
	K = linear stiffness matrix of structure; P = outer load; V = node displacements			

Table 3.2

Formation of additional load for the allowance of the main nonlinear properties of reinforced concrete on the basis of its additional design diagrams

№	General order of formation of additional load	Plastic properties of concrete	Nonlinearity of bond between concrete and reinforcement	Cracking	Action of temperature	Prestressing	Reload and unload
1	$K_{nonl}\,V = P$	$K_{pe}\,V = P$	$K_{bond}\,V = P$	$K_{crc}\,V = P$	$K_t\,V = P$	$K_{pr}\,V = P$	$K_{rep}\,V = P$
2	$K_{nonl} =$ $= K+\Delta K_{nonl}$	$K_{pl} =$ $= K + \Delta K_{pl}$	$K_{bond} =$ $= K + \Delta K_{bond}$	$K_{crc} =$ $= K + \Delta K_{crc}$	$K_t =$ $= K_e + \Delta K$	$K_{pr} =$ $= K + \Delta K_{pr}$	$K_{rep} =$ $= K + \Delta K_{rep}$
3	$(K+\Delta K_{nonl})V=$ $=P$	$(K+\Delta K_{pl})V=$ $=P$	$(K+\Delta K_{bond})V=$ $=P$	$(K+\Delta K_{crc})V=$ $=P$	$(K+\Delta K_t)V=$ $=P$	$(K+\Delta K_{pr})V=$ $=P$	$(K+\Delta K_{rep})V=$ $=P$
4	$K\,V=$ $=P - \Delta K_{nonl}V$	$K\,V=$ $=P - \Delta K_{pl}\,V$	$K\,V=$ $=P - \Delta K_{bond}\,V$	$K\,V=$ $=P - \Delta K_{crc}\,V$	$K\,V=$ $=P - \Delta K_t V$	$K\,V=$ $=P - \Delta K_{pr}\,V$	$K\,V=$ $=P - \Delta K_{rep}\,V$
5	$F_{nonl} =$ $= - \Delta K_{nonl}V$	$F_{pl} =$ $= - \Delta K_{pl}\,V$	$F_{bond} =$ $= - \Delta K_{bond}V$	$F_{crc} =$ $= -\Delta K_{crc}\,V$	$F_t =$ $= -\Delta K_t V$	$F_{pr} =$ $= - \Delta K_{pr}\,V$	$F_{rep} =$ $= - \Delta K_{rep}\,V$

3.1.2. Simplified order of formation of additional load on the basis of additional finite elements

The simplified algorithm of formation of additional loads by additional finite elements properties which allows to avoid composition of additional design diagram of structure was given in section 1.5. In this case the vector of additional load of the whole structure is formed from the vectors of additional load for each finite element of the design diagram. In formation of such vector for the allowance of any nonlinearity manifested by the definite finite element as its ultimate limit state is reaching next operations are used:

1. Representation of relationship between node reactions and node displacements for each finite element for the allowance of the *i*-th nonlinear property in the form of: $K_{nonl,e,i}\ V = R_{nonl,e,i}$ (see 1.23).

2. Consideration of stiffness matrix of finite element with the *i*-th nonlinear property as the sum of its stiffness matrix $K_{nonl,e,i-1}$ without this property and stiffness matrix of additional finite element $\Delta K_{nonl,e,i}$ destined to take into account this nonlinear property: $K_{nonl,e,i} = K_{nonl,e,i-1} + \Delta K_{nonl,e,i}$ (see 1.24).

3. Calculation of a stiffness matrix of the additional finite element: $\Delta K_{nonl,e,i} = K_{nonl,e,i} - K_{nonl,e,i-1}$.

4. Calculation of node reactions $\Delta R_{nonl,i}$ of additional finite element according to general formula: $\Delta R_{nonl,i} = \Delta K_{nonl,e,i}\ V$ (see 1.27). Besides this formula which is identical for both types of additional finite elements it is possible to use others characteristic for each of these two types. For an additional finite element of the first type it is the relationship between its node reactions $\Delta R_{nonl,i}$ and stresses $\Delta\sigma_{nonl,i}$: $\Delta R_{nonl,i} = C\Delta\sigma_{nonl,i}$ (see 1.41). For an additional finite element of the second type it is the relationship between its node reactions $\Delta R_{nonl,i}$ and strains $\Delta\varepsilon_{nonl,i}$: $\Delta R_{nonl,i} = G\ \Delta\varepsilon_{nonl,i}$ (see 1.55).

5. Determination of an additional load for the allowance of the *i*-th nonlinear property of the definite finite element: $F_{e,i} = -\ \Delta R_{nonl,i}$ (see 1.28).

The sequence of formation of the vector of additional load for the allowance of the main nonlinear properties of reinforced concrete structures for the separate finite element from design diagram is given in Table 3.4. List of the taken into account nonlinear properties and notation of additional load corresponds to the represented ones in Table In.1 and Table 3.2. The system of notations is given in Table 3.3.

The order of formation of additional load for the allowance of the main nonlinear properties of reinforced concrete structures corresponds to the simplified algorithm described in section 1.5. This algorithm is based on use of the properties of additional finite elements for the allowance of each definite nonlinear property as the ultimate limit state is reaching and allows to avoid the operation of composition of additional design diagram for the whole structure since this design diagram repeats geometry of the main design diagram.

Table 3.3

Necessary values for formation of additional load of the separate finite element on the basis of its additional finite elements

№	General title of value	General notation in nonlinearity	Plastic properties of concrete	Nonlinearity of bond between concrete and reinforcement	Cracking	Reload and unload	Action of temperature	Pre-stressing
1	Stiffness matrix of finite element with nonlinear properties	$K_{nonl,e,i}$	$K_{pl,e}$	$K_{bond,e}$	$K_{crc,e}$	$K_{rep,e}$	$K_{t,e}$	$K_{pr,e}$
2	Stiffness matrix of additional finite element	$\Delta K_{nonl,e,i}$	$\Delta K_{pl,e}$	$\Delta K_{bond,e}$	$\Delta K_{crc,e}$	$\Delta K_{rep,e}$	$\Delta K_{t,e}$	$\Delta K_{pr,e}$
3	Node reactions of additional finite element	$\Delta R_{nonl,i}$	ΔR_{pl}	ΔR_{bond}	ΔR_{crc}	ΔR_{rep}	ΔR_t	ΔR_{pr}
4	Stresses in additional finite element	$\Delta\sigma_{nonl,i}$	$\Delta\sigma_{pl}$	$\Delta\sigma_{bond}$	$\Delta\sigma_{crc}$	$\Delta\sigma_{rep}$	$\Delta\sigma_t$	$\Delta\sigma_{pr}$
5	Strains in additional finite element	$\Delta\varepsilon_{nonl,i}$	$\Delta\varepsilon_{pl}$	$\Delta\varepsilon_{bond}$	$\Delta\varepsilon_{crc}$	$\Delta\varepsilon_{rep}$	$\Delta\varepsilon_t$	$\Delta\varepsilon_{pr}$
6	Additional load	$F_{e,i}$	$F_{pl,e}$	$F_{bond,e}$	$F_{crc,e}$	$F_{rep,e}$	$F_{t,e}$	$F_{pr,e}$
7	Other values	K_e = stiffness matrix of finite element without considered nonlinear property; V = node displacement of finite element						

The properties of additional finite elements ought to be determined before the formation of additional load for the whole structure F_{nonl}. The additional load $F_{e,i}$ of each the main finite elements of design diagram may be calculated by these additional finite elements only. The additional load of the whole structure F_{nonl} is formed on the basis of the values $F_{e,i}$.

Such algorithm presupposes that additional blocks of program which allow to determine the properties of additional finite elements ought to be created for definition of additional load. At the same time the development and the inclusion in program of such additional blocks are not a serious problem because it does not affect the main algorithm of problem solving. It means that two ways may be used for realization of nonlinear analysis at limit state: creation of the new nonlinear programs and development of additional programs for previously created linear ones.

Table 3.4

Formation of additional load for the allowance of the main nonlinear properties of reinforced concrete for separate finite element on the basis of its additional finite elements

№	General order of formation of additional load		Plastic properties of concrete	Nonlinearity of bond between concrete and reinforcement	Cracking	Prestressing	Reload and unload	Action of temperature
1	$K_{nonl,e,}V = R_{nonl,e,i}$		$K_{pl,e}V = R_{pl}$	$K_{bond,e}V = R_{bond}$	$K_{crc,e}V = R_{crc}$	$K_{pr,e}V = R_{pr}$	$K_{rep,e}V = R_{rep}$	$K_{t,e}V = R_t$
2	$K_{nonl,e,i} = K_{nonl,e,i-1} + \Delta K_{nonl,e,i}$		$K_{pl,e} = K_e + \Delta K_{pl,e}$	$K_{bond,e} = K_e + \Delta K_{bond,e}$	$K_{crc,e} = K_e + \Delta K_{crc,e}$	$K_{pr,e} = K_e + \Delta K_{pr,e}$	$K_{rep,e} = K_e + \Delta K_{rep,e}$	$K_{t,e} = K_e + \Delta K_{t,e}$
3	$\Delta K_{nonl,e,i} = K_{nonl,e,i} - K_{nonl,e,i1}$		$\Delta K_{pl,e} = K_{pl,e} - K_e$	$\Delta K_{bond,e} = K_{bond,e} - K_e$	$\Delta K_{crc,e} = K_{crc,e} - K_e$	$K_{pr,e} = K_{pr,e} - K_e$	$\Delta K_{rep,e} = K_{rep,e} - K_e$	$\Delta K_{t,e} = K_{t,e} - K_e$
4	$\Delta R_{nonl,i} = \Delta K_{nonl,e,}V$		$\Delta R_{pl} = \Delta K_{pl,e}V$	$\Delta R_{bond} = \Delta K_{bond,e}V$	$\Delta R_{crc} = \Delta K_{crc,e}V$	$\Delta R_{pr} = \Delta K_{pr,e}V$	$\Delta R_{rep} = \Delta K_{rep,e}V$	$\Delta R_t = \Delta K_{t,e}V$
	The1-th type	$\Delta R_{nonl,i} = C\Delta\sigma_{nonl}$	$\Delta R_{pl} = C\Delta\sigma_{pl}$	$\Delta R_{bond} = C\Delta\sigma_{bond}$	$\Delta R_{crc} = C\Delta\sigma_{crc}$	$\Delta R_{pr} = C\Delta\sigma_{pr}$	$\Delta R_{rep} = C\Delta\sigma_{rep}$	$\Delta R_t = C\,\Delta\sigma_t$
	The 2-d type	$\Delta R_{nonl,i} = G\Delta\varepsilon_{nonl}$	$\Delta R_{pl} = G\Delta\varepsilon_{pl}$	$\Delta R_{bond} = G\Delta\varepsilon_{bond}$	$\Delta R_{crc} = G\Delta\varepsilon_{crc}$	$\Delta R_{pr} = G\Delta\varepsilon_{pr}$	$\Delta R_{rep} = G\,\Delta\varepsilon_{rep}$	$\Delta R_t = G\Delta\varepsilon_t$
5	$F_{e,i} = -\Delta R_{nonl,i}$		$F_{pl,e} = -\Delta R_{pl}$	$F_{bond,e} = -\Delta R_{bond}$	$F_{crc,e} = -\Delta R_{crc}$	$F_{pr,e} = -\Delta R_{pr}$	$F_{rep,e} = -\Delta R_{rep}$	$F_{t,e} = -\Delta R_t$

3.1.3. Formation of additional loads at limit state

The problem of formation of additional loads for analysis of structures at ultimate limit state requires a special attention due to its theoretical significance.

As it was shown in section 1.2.1 the ultimate limit state of structure appears after reaching of ultimate limit state in several finite elements the number of which is critical for the whole structure.

It means that additional load for the allowance of the ultimate limit state ought to be formed gradually in accordance with the number of those finite elements in which their characteristic ultimate limit states are reached. Therefore the main problem is the question of way of formation of additional load for each of the main finite elements depending on the character of their ultimate limit state.

The main idea for solving of this problem by additional finite elements was shown in section 1.2.1. In this section it should be concentrated on particular calculating operations of realization of problem solving.

Before the reaching of ultimate limit state the main finite element exhibits all its characteristic nonlinear properties therefore the properties of additional finite element are formed in accordance with this fact. It means that the formation of additional load before the reaching of ultimate limit state remains unchanged and corresponds to the data of the previous section. After the reaching of limit state the main finite element may have partial or complete collapse.

Complete collapse means that this element does not influence behavior of the neighboring ones and its node reactions are equal to 0 ($R_{nonl,e,lim}$ = 0). The properties of additional finite element are formed by corresponding way in accordance with design diagram given in Fig. 1.13. These properties are determined by the properties of the main linear finite element. The additional load $F_{e,lim}$ is determined by node reactions of linear finite element too.

Partial collapse means that the main finite element remains partial influence on behavior of the structure and its node reactions are not equal to 0 ($R_{nonl,e,lim} \neq$ 0). First the partial unload of this element due to taking off ultimate limit stresses takes place and then the partial reload happens.

The sequence of formation of additional load for the allowance of the ultimate limit state of the separate finite element of design diagram depending on its complete or partial collapse was given in Table 3.5. The system of notations is presented in Table 3.6.

Table 3.5

Formation of additional load for the allowance of the ultimate limit state of separate finite element on the basis of additional finite elements

№	General order of formation of additional load at ultimate limit state	Two ways of collapse of finite element after reaching of ultimate limit state		
		Partial collapse		Collapse
		Unload	Reload	
1	$K_{nonl,e,lim}V=R_{nonl,e,lim}$	$K_{dis,e,lim}V=R_{dis,e,lim}$	$K_{rep,e,lim}V=R_{rep,e,lim}$	$R_{nonl,e,lim} = 0$ $K_{nonl,e,lim}\ V= 0$
2	$K_{nonl,e,lim}=$ $= K_e + \Delta K_{nonl,e,lim}$	$K_{dis,e,lim}=$ $= K_e + \Delta K_{dis,e,lim}$	$K_{rep,e,lim}=$ $= K_e + \Delta K_{rep,e,lim}$	$K_{nonl,e,lim} = 0$
3	$\Delta K_{nonl,e,lim} =$ $= K_{nonl,e,lim} - K_e$	$\Delta K_{dis,e,lim} =$ $= K_{dis,e,lim} - K_e$	$\Delta K_{rep,e,lim} =$ $= K_{rep,e,lim} - K_e$	$\Delta K_{nonl,e,lim}= -K_e$
4	$\Delta R_{nonl,e,lim} =$ $= \Delta K_{nonl,e,lim}\ V$	$\Delta R_{dis,e,lim} =$ $= \Delta K_{dis,e,lim}\ V$	$\Delta R_{rep,e,lim} =$ $= \Delta K_{rep,e,lim}\ V$	$\Delta R_{nonl,e,lim} =$ $= -K\ V$
5	$F_{e,lim}= - \Delta R_{nonl,e,lim}$	$F_{dis,e,lim}= -\Delta R_{dis,e,li}$	$F_{rep,e,lim}=-\Delta R_{rep,e,lim}$	$F_{e,lim}= R$

The sequence of formation of additional load for the allowance of the ultimate limit state is based on use of properties of additional finite elements for the allowance of the complete or partial loss by the main finite element of its bearing capacity at reaching of ultimate limit state in it.

The properties of additional finite elements for the allowance of the behavior of the main finite element at ultimate limit state and its partial collapse after it ought to be determined before the moment of formation of additional load as the additional load $F_{e,lim}$ for each main finite element of design diagram from which the complete additional load F_{nonl} of the whole structure is formed, may be calculated on the basis of these properties only.

With the exception of additional finite elements for the allowance of collapse of the main finite element as their properties since the additional load $F_{e,lim}$ are determined by the properties of linear finite element.

Algorithm of formation of additional loads at ultimate limit state is a special case of algorithm of their formation by additional finite elements. The properties of additional finite elements for the allowance of the ultimate limit state of the corresponding main finite elements are used in this case.

Table 3.6

Necessary values for formation of additional load for the allowance of the ultimate limit state of separate finite element

№	General title of value	Ultimate limit state or collapse of finite element	Partial collapse of finite element	
			Unload	Reload
1	Stiffness matrix of finite element	$K_{nonl,e,lim}$	$K_{dis,e,lim}$	$K_{rep,e,lim}$
2	Stiffness matrix of additional finite element	$\Delta K_{nonl,e,lim}$	$\Delta K_{dis,e,lim}$	$\Delta K_{rep,e,lim}$
3	Node reactions of additional finite element	$\Delta R_{nonl,e,lim}$	$\Delta R_{dis,e,lim}$	$\Delta R_{rep,e,lim}$
4	Additional load	$F_{e,lim}$	$F_{dis,e,lim}$	$F_{rep,e,lim}$
5	Other values	K_e = stiffness matrix of linear finite element; R = node reactions of linear finite element; V = node displacements		

It means that the main sequence of problem solving remains unchanged in analysis at ultimate limit state. The realization of this analysis requires only the auxiliary blocks which do not change the general structure of using programs.

3.2. Additional triangular concrete deep-beam finite element for the allowance of the nonlinear properties

Next theoretical elaborations are in the basis of mathematical description of the properties of additional triangular deep-beam finite element with allowance for nonlinear properties: triangular deep-beam finite element with elastic properties (see Section In.1.3); triangular deep-beam finite element with conditional crack (see Section In.3.1); way of formation additional load similar to developed by Karjakin A.A. on the basis deformation theory of plasticity created by Geniev G.A.; general order of formation of additional finite elements properties is stated above. The first step of this way ought to be the study of a procedure of formation for properties of the considered additional finite element depending on the stage of behavior of analogous triangular concrete finite element.

The stages of behavior of triangular concrete deep-beam finite element which it goes through before the reaching of its two typical ultimate limit states in compression and tension were considered in section 1.2.3.

It means that the properties of corresponding additional finite element ought to be formed depending on the nonlinear properties which are exhibited at each definite stage of its behavior. The specific examples illustrating such approach are given later.

3.2.1. Additional triangular concrete deep-beam finite element for operating at ultimate compression

As it was shown in section 1.2.3 the triangular concrete deep-beam finite element in ultimate limit compression goes through two stages of behavior: stage of plastic behavior before reaching of ultimate limit state and collapse after its reaching.

It means that the properties of corresponding additional finite element are determined by these two stages too. At first it is a triangular deep-beam additional finite element for the allowance of the plastic properties, then the same element providing collapse of the main finite element.

The properties of additional finite element for the allowance of the plastic properties of concrete are presented in Table 3.7. It is better to use a simpler way for determination of coefficient of plasticity ω instead of way described in s. In.3.1. For example, $\omega = 1-\varepsilon_{b,ult}{}^{2}/((\varepsilon-\varepsilon_{b,ult})^{2}+E_{b}\varepsilon_{b,ult}{}^{2}\varepsilon/R_{b})$, where $\varepsilon_{b,ult}$ is ultimate strain of concrete in compression. In tension we take R_{bt} and $\varepsilon_{bt,ult}$ in place of R_b and $\varepsilon_{b,ult}$.

Table 3.7

The main characteristics of triangular deep-beam additional finite element for the allowance of the plastic properties of concrete

№	Type of characteristic	Finite element with linear properties	Finite element with plastic properties	Additional finite element for the allowance of the plastic properties
1	Relationship between node reactions and node displacements	$R = K_e V$	$R_{pl} = K_{e,pl} V$	$\Delta R_{pl} = \Delta K_{e,pl} V$
2	Stiffness matrix	$K_e = (E_b/(1-\nu^2))tS(A^{-1})^T B^T D B A^{-1}$	$K_{e,pl} = (1-\omega)K_e$	$\Delta K_{e,pl} = -\omega K$
3	Relationship between node reactions and stresses	$R = t\,S(A^{-1})^T B\,\sigma$	$R_{pl} = tSC\sigma_{pl}$	$\Delta R_{pl} = t\,S\,(A^{-1})^T B^T \Delta\sigma_{pl}$
4	Relationship between strains and displacements	$\varepsilon = B\,A^{-1}\,V$	$\varepsilon_{pl} = \varepsilon$	$\Delta\varepsilon_{pl} = 0$ (The 1 type of AFE)
5	Relationship between stresses and strains	$\sigma = \lambda D\varepsilon$	$\sigma_{pl} = \lambda D\varepsilon + \Delta\sigma_{pl}$	$\Delta\sigma_{pl} = -\omega\,\lambda D\varepsilon$
6	Additional load	$F_{e,pl} = -\Delta R_{pl}$	–	–

Note that the additional finite element changing stress state of the main finite element, i.e. additional finite element of the first type (see Section 1.5.2) is used in this case. This type of additional finite element is a preferred one in the case as it is simple to obtain the required mathematical relationships by means of previously made theoretical elaborations.

The properties of triangular additional finite element providing collapse of the main finite element after reaching of ultimate limit state in compression are given in Table 3.8. These properties are completely determined by characteristics of the main finite element as they ought to eliminate the influence of the main finite element on the neighboring ones. As in previous case the additional finite element of the first type is used here because the obtaining of mathematical relationships is rather simple.

These two additional finite elements allow completely describe behavior of the main concrete triangular deep-beam finite element in compression. In view of the fact on the basis of their properties we may form the additional load of the main linear finite element depending on plastic properties $F_{e,pl}$ and collapse $F_{e,lim}$ after reaching of the ultimate limit compression stage.

On the basis of these vectors it is possible to form additional load for all finite elements of design diagram of the considered structure at ultimate limit compression.

Table 3.8

The main characteristics of triangular deep-beam additional finite element for the allowance of the ultimate limit state

№	Type of characteristic	Finite element with linear properties	Finite element at ultimate limit state	Additional finite element for account of ultimate limit stat
1	Relationship between node reactions and node displacements	$R = K_e\, V$	$R_{e,lim} = 0$	$\Delta R_{e,lim} = \Delta K_{e,lim}\, V$
2	Stiffness matrix	$K_e = (E_b/(1-\nu^2))tS(A^{-1})^T B^T D B A^{-1}$	$K_{e,lim} = 0$	$\Delta K_{e,lim} = -K_e$
3	Relationship between node reactions and stresses	$R = tS(A^{-1})^T B^T \sigma$	$R_{e,lim} = 0$	$\Delta R_{e,lim} = tS(A^{-1})^T B^T \Delta\sigma$
4	Relationship between strains and displacements	$\varepsilon = B\, A^{-1}\, V$	$\varepsilon_{e,lim} = \varepsilon$	$\Delta\varepsilon_{e,lim} = 0$ (The 1 type of AFE)
5	Relationship between stresses and strains	$\sigma = \lambda\, D\, \varepsilon$	$\sigma_{e,lim} = 0$	$\Delta\sigma_{e,lim} = -\sigma$
6	Additional load	$F_{e,lim} = R$	–	–

3.2.2. Additional triangular concrete deep-beam finite element for operating at ultimate limit tension

Concrete deep-beam finite element at ultimate limit tension (see Section 1.2.3) goes through four stages: plastic behavior at tension before crack; partial unload due to cracking; partial plastic behavior after cracking before reaching of ultimate limit state at reload; collapse after reaching of ultimate limit state at reload. Two of these four stages namely the first stage (plastic behavior after initial load) and the fourth stage (collapse) coincide with the first stage and the second stage respectively of behavior of triangular concrete finite element at ultimate limit compression. It means that analytic relations of the used additional finite element of the first type coincide too and are represented earlier in the previous section in Table 3.7 and Table 3.8.

As to remainder the second stage and the third one there are corresponding the additional finite elements destined for the allowance of the partial unload due to crack and the presence of residual strains in concrete in view of cracking and the additional finite element for the allowance of the behavior of element with crack. The properties of the first additional finite element of these two ones are presented in Table 3.9. Additional finite element of the second type (see Section 1.5.3), i.e. the additional finite element changing strain state of the main finite element was used in development of these properties.

Table 3.9

The main characteristics of triangular deep-beam additional finite element for the allowance of the partial unload

№	Type of characteristic	Finite element with linear properties	Finite element for the allowance of the partial unload	Additional finite element for the allowance of the partial unload
1	Relationship between node reactions and node displacements	$R = K_e\, V$	$R_{dis,e,lim} = K_{dis,e,lim}\, V$	$\Delta R_{dis,e,lim} = \Delta K_{e,lim}\, V$
2	Stiffness matrix	$K= =\lambda tS(A^{-1})^T B^T DBA^{-1}$	$K_{dis,e,lim} = K_{crc}$	$\Delta K_{dis,e,lim}=K_{dis,e,lim} - K_e$
3	Relationship between node reactions and stresses	$R = =tS\lambda(A^{-1})^T B^T D\varepsilon$	$R_{dis,e,lim}= =tS(A^{-1})^T B^T\, D\varepsilon_{dis,e,lim}$	$\Delta R_{dis,e,lim}= =tS(A^{-1})^T B^T D\Delta\varepsilon_{dis,e,lim}$
4	Relationship between strains and displacements	$\varepsilon =BA^{-1}V$	$\varepsilon_{dis,e,lim} = =0{,}9\ (\varepsilon - D^{-1}\sigma_{pl,lim}/\lambda)$	$\Delta\varepsilon_{dis,e,lim} = =\varepsilon - 0{,}9(\varepsilon - D^{-1}\sigma_{pl,lim}/\lambda)$
5	Relationship between stresses and strains	$\varepsilon = D^{-1}\sigma/\lambda$	$\varepsilon_{dis,e,lim} = = 0{,}9(\varepsilon - D^{-1}\sigma_{pl,lim}/\lambda)$	$\Delta\sigma_{dis,e,lim} = 0$ (The 2 type of AFE)
6	Additional load	$F_{dis,e,lim} = = -\Delta R_{dis,e,lim}$	–	–

In this case the second type of additional finite element is preferred as it is simple to take into account the main property of concrete after unload namely the presence of residual strains. If before the cracking there were stresses $\sigma_{pl,lim}$ in concrete finite element then residual strains $\varepsilon_{dis,e,lim}$ may be determined by elastic strains ε according next formula: $\varepsilon_{dis,e,lim} = \varepsilon_{pl,2} = \varepsilon_{pl} - 0{,}1\varepsilon_{pl} = 0{,}9\varepsilon_{pl} = 0{,}9\ (\varepsilon - D^{-1}\sigma_{pl,lim}/\lambda)$.

The properties of an additional finite element for the allowance of the crack in the main finite element are given in Table 3.10. The triangular finite element with conditional crack which properties were given in section In.3.1 is in the basis of this additional finite element. Two these additional finite elements together with two those given in the previous section allow to describe completely the behavior of triangular concrete deep-beam finite element at ultimate limit tension. On the basis of properties of each of these four finite elements the additional load for the main linear finite element with allowance for its real behavior depending on particular stage may be formed. Additional load for the whole structure may be formed from vectors of additional load of separate finite elements at ultimate limit tension.

Table 3.10

The main characteristics of triangular deep-beam additional finite element for the allowance of the conditional crack

№	Type of characteristic	Finite element with linear properties	Finite element with conditional crack	Additional finite element for the allowance of the conditional crack
1	Relationship between node reactions and node displacements	$R = K\,V$	$R_{crc} = K_{crc}\,V$	$\Delta R_{crc} = \Delta K_{crc}\,V$
2	Stiffness matrix	$K =$ $=\lambda tS(A^{-1})^T B^T DBA^{-1}$	$K_{crc}=T^T K'T$	$\Delta K_{crc}=K_{crc} - K$
3	Relationship between node reactions and stresses	$R = tS(A^{-1})^T B^T\,\sigma$	$R_{crc} =$ $= tS(A^{-1})^T B^T\sigma_{crc}$	$\Delta R_{crc} =$ $= tS(A^{-1})^T B^T \Delta\sigma_{crc}$
4	Relationship between strains and displacements	$\varepsilon = BA^{-1}V$	$\varepsilon_{crc}=\varepsilon$	$\Delta\varepsilon_{crc} = 0$ (The 1 type of AFE)
5	Relationship between stresses and strains	$\sigma = \lambda D\varepsilon$	$\sigma_{crc} = a_\alpha\sigma_{mc}$	$\Delta\sigma_{crc} = \sigma_{crc} - \sigma$
6	Additional load	$F_{e,crc}=$ $= -\Delta R_{crc}$	–	–

3.2.3. Example of determination of stiffness matrix of triangular concrete deep-beam additional finite element

As it was indicated earlier a stiffness matrix is the main characteristic determining the properties of the main finite element and the additional one. This stiffness matrix ought to be determined depending on the stage of behavior of the main finite element. Example of its calculation for right triangle finite element is given in Table 3.11. Next variants of the behavior of the main finite element are considered there: linear (initial) behavior, plastic behavior, two variants of behavior with crack (with preliminary linear behavior of concrete and plastic one) and collapse.

Table 3.11

Example of calculation of stiffness matrix of triangular concrete deep-beam additional finite element depending on the stage of its behavior

№	Stiffness matrix of finite element with linear properties K_e and its structure	Stiffness matrix of finite element with nonlinear properties $K_{nonl,e} = K_e + \Delta K_{nonl,e}$	Stiffness matrix of additional finite element $\Delta K_{nonl,e} = K_{nonl,e} - K_e$
1	2	3	4
1	1.Shape of finite element and structure of its stiffness matrix Y(u) k a i j 0 X(w) a E_b = 300000 kg/cm^2; a = 5,0 cm; t = 1,0 cm	K_{uiui} K_{uiwi} K_{uiuj} K_{uiwj} K_{uiuk} K_{uiwk} K_{wiui} K_{wiwi} K_{wiuj} K_{wiwj} K_{wiuk} K_{wiwk} K_{ujui} K_{ujwi} K_{ujuj} K_{ujwj} K_{ujuk} K_{ujwk} K_{wjui} K_{wjwi} K_{wjuj} K_{wjwj} K_{wjuk} K_{wjwk} K_{ukui} K_{ukwi} K_{ukuj} K_{ukwj} K_{ukuk} K_{ukwk} K_{wkui} K_{wkwi} K_{wkuj} K_{wkwj} K_{wkuk} K_{wkwk}	
2	2.Linear analysis ($K_{nonl,e} = K_e$, $\Delta K_{nonl,e} = 0$)		
	218750 93750 –62500 –31250 –156250 –62500 218750 –62500 –156250 –31250 –62500 62500 0 0 62500 156250 31250 0 *Symmetry* 156250 0 62500	218750 93750 –62500 –31250 –156250 –62500 218750 –62500 –156250 –31250 –62500 62500 0 0 62500 156250 31250 0 *Symmetry* 156250 0 62500	0 0 0 0 0 0 0 0 0 0 0 0 0 0 0 0 0 0 *Symmetry* 0 0 0
3	3. Plastic analysis where ω = 0,5 ($K_{pl,e} = K_e + \Delta K_{pl,e}$)		
	18750 93750 –62500 –31250 –156250 –62500 218750 –62500 –156250 –31250 –62500 62500 0 0 62500 156250 31250 0 *Symmetry* 156250 0 62500	09375 46875 –31250 –15625 –78125 –31250 109375 –31250 –78125 –15625 –31250 31250 0 0 31250 78125 15625 0 *Symmetry* 78125 0 31250	109375 –46875 31250 15625 78125 31250 –109375 31250 78125 15625 31250 –31250 0 0 –31250 –78125 –15625 0 *Symmetry* –78125 0 –31250
4	**4.** Behavior with crack without account of plastic properties where α = 30° and ω = 0 ($K_{crc,e} = K_e + \Delta K_{crc,e}$)		
	18750 93750 –62500 –31250 –156250 –62500 218750 –62500 –156250 –31250 –62500 62500 0 0 62500 156250 31250 0 *Symmetry* 156250 0 62500	09930 121200 –76840 –44362 –133088 –76838 69976 –44364 –25614 –76838 –44364 28126 16238 48714 28126 9376 28126 16238 *Symmetry* 84374 48714 28126	–8820 27450 –14340 –13112 23162 –14338 –148774 18136 130636 –45588 18136 –34374 16238 48714 –34374 –146874 –3124 16238 *Symmetry* –71876 48714 –34374
5	5. Behavior with crack with account of plastic properties where α = 30° and ω = 0,5 ($K_{crc,e} = K_{pl,e} + \Delta K_{crc,e}$)		
	09375 46875 –31250 –15625 –78125 –31250 109375 –31250 –78125 –15625 –31250 31250 0 0 31250 78125 15625 0 *Symmetry* 78125 0 31250	04965 60600 –38420 –22182 –66544 –39419 34988 –22182 –12807 –38419 –22182 14063 8119 24357 14063 4688 14063 8119 *Symmetry* 42063 24314 14063	–4410 13725 –7170 –6556 11581 –7160 –74387 9068 65318 –22794 9068 –17187 8119 24357 –17187 –73437 –1562 8119 *Symmetry* –35938 24357 –17187

1	2	3	4
6	6. Collapse ($K_{lim,e} = 0$, $\Delta K_{lim,e} = -K_e$)		
	$\begin{array}{rrrrrr} 18750 & 93750 & -62500 & -31250 & -156250 & -62500 \\ & 218750 & -62500 & -156250 & -31250 & -62500 \\ & & 62500 & 0 & 0 & 62500 \\ & & & 156250 & 31250 & 0 \\ \textit{Symmetry} & & & & 156250 & 0 \\ & & & & & 62500 \end{array}$	$\begin{array}{rrrrrr} 0 & 0 & 0 & 0 & 0 & 0 \\ & 0 & 0 & 0 & 0 & 0 \\ & & 0 & 0 & 0 & 0 \\ & & & 0 & 0 & 0 \\ \textit{Symmetry} & & & & 0 & 0 \\ & & & & & 0 \end{array}$	$\begin{array}{rrrrrr} 218750 & -93750 & 62500 & 31250 & 156250 & 62500 \\ & -218750 & 62500 & 156250 & 31250 & 62500 \\ & & -62500 & 0 & 0 & -62500 \\ & & & -156250 & -31250 & 0 \\ \textit{Symmetry} & & & & -156250 & 0 \\ & & & & & -62500 \end{array}$

3.3. Additional connecting finite element for the allowance of the nonlinearity of bond between concrete and reinforcement

If we consider behavior of connecting finite element which simulates bond (see Section In.1.4 and Table 3.12) we may isolate two ways of ultimate limit state reaching which are distinguished by behavior of contact between concrete and reinforcement.

The first way is characteristic for conditions of normal (transverse) pressure of concrete on reinforcement and consists of two stages: 1) behavior at presence of longitudinal bond along reinforcement under normal bond before reaching of ultimate limit state; 2) at absence of bond after reaching of ultimate limit state. In mathematics these stages correspond to presence of longitudinal and transverse bonds in the connecting finite element and absence of these bonds.

The second way is possible in case of disappearance of transverse pressure of concrete on reinforcement in the process of their interaction and its realization consists of three stages: 1) initial behavior in the presence of longitudinal bond along reinforcement under transverse pressure before the moment of its disappearance; 2) behavior in the presence of longitudinal bond in absence of transverse pressure before reaching of ultimate limit state; 3) absence of bond after reaching of ultimate limit state. In mathematics realization of the second way means at first the presence of longitudinal and transverse bonds then the presence of longitudinal bonds in the absence of transverse ones and finally in the absence of both types of bond.

Additional connecting finite element for the allowance of the nonlinearity of bond between concrete and reinforcement ought to reflect both possible variants of behavior of the main connecting finite element. Its characteristics are presented in Table 3.12.

In the basis of additional connecting finite element are the properties of connecting element with linear properties and differentiated law of bond between concrete and reinforcement stated by Oatul A.A. Additional load for the whole structure is formed from vectors of additional load $F_{e,bond}$ for the allowance of the nonlinear properties of bond between concrete and reinforcement concrete of each main connecting finite element.

Table 3.12

The main characteristics of additional connecting finite element for the allowance of the nonlinear properties of bond between concrete and reinforcement

	Type of characteristic	Finite element with linear properties	Finite element with nonlinearity of bond	Additional connecting finite element for the allowance of the nonlinearity of bond
1	Relationship between node reactions and node displacements	$R = K_e V$	$R_{bond} = K_{e,bond} V$	$\Delta R_{bond} = \Delta K_{e,bond} V$
2	Stiffness matrix a) presence of longitudinal and transverse bond b) absence of transverse bond c) ultimate limit state (collapse)	$K_e = f(K_x, K_y)$ $K_x = S_a, A_1$ $K_y = S_a, A_1, tg\varphi$ Section a Section a	$K_{e,ond} = f(K_x, K_y)$ $K_x = f(S_a, A_1, g)$ $K_y = f(S_a, A_1, tg\varphi, g)$ $K_x = f(S_a, A_1, g)$ $K_y = 0$ $K_e = 0$ $K_x = K_y = 0$	$\Delta K_{e,bond} = K_{e,bond} - K_e$ Section a $\Delta K_{e,bond} = -K_e$
3	Relationship between node reactions and stresses	$\tau = R/S_a$	$\tau_{bond} = R_{bond}/S_a$	$\Delta\tau_{bond} = \Delta R_{bond}/S_a$
4	Relationship between mutual shifts and displacements	$g = V_b - V_s$	$g_{bond} = g$	$\Delta g_{bond} = 0$ (The 1 type of additional connecting FE)
5	Relationship between stresses of bond and mutual shifts of nodes	$\tau = f(A_1, tg\varphi, g)$	$\tau_{bond} = f(A_1, tg\varphi, C, g_{bond})$	$\Delta\tau_{bond} = \tau_{bond} - \tau$
6	Additional load	$F_{e,bond} = -\Delta R_{bond}$	–	–

3.4. Additional loads for nonlinear analysis of reinforced concrete structures operating in plane stress state

Examples of formation of additional loads on the basis of additional triangular finite elements and connecting ones for the allowance of the different properties of reinforced concrete structures at plane stress state were considered in previous sections. It allows to generalize the obtained theoretical data in Table 3.13 and use them for development of algorithm and programs. This table allows to determine additional loads and correcting vectors of stresses or strains on the basis of properties of additional finite element of the general, first and second types are destined for the allowance of the nonlinear property.

In this case design diagram of the considered structure may consist of two types of the main finite elements: triangular deep-beam finite elements and two-node connecting finite elements. The first type of finite elements may be used for simulation of concrete and reinforcement. The second type may be used for simulation of bond between concrete and reinforcement. At the same time the allowance for nonlinear properties is carried out for concrete finite elements since SN and Ps forbid operating of reinforcement at plastic stage. Thus the stated method allows to carry out nonlinear analysis of reinforced concrete structures at plane stress state at all stages of behavior from the beginning of load till reaching of ultimate limit state. Simultaneously this procedures is the prove of efficiency of formation of additional loads and correcting vectors on the basis of properties of additional finite elements and opens opportunity for analogous development in the field of nonlinear analysis of reinforced concrete structures at other stress states.

Table 3.13

Additional loads and correcting vectors of stresses and strains for nonlinear finite elements and connecting finite elements in analysis of reinforced concrete structures at plane stress state

№	Nonlinear property	Theoretical basis	Additional load		Correcting vector		Note
			Formula	Defining value	Notation	Title	
1	2	3	4	5	6	7	8
1	Plastic properties of concrete	Additional triangular deep-beam finite element (AFE) of the first type for the allowance of the plastic properties of concrete	$F_{e,pl} = -\Delta R_{pl}$	Node reactions in AFE	$\Delta\sigma_{pl}$	Stresses in AFE	Table 3.7
2	Cracking of concrete	Additional triangular deep-beam finite element of the first type for the allowance of the conditional crack	$F_{e,crc} = -\Delta R_{crc}$	Node reactions in AFE	$\Delta\sigma_{crc}$	Stresses in AFE	Table 3.10
3	Bond between concrete and reinforcement	Additional connecting element (ACE) of general type for the allowance of the nonlinear properties of bond between concrete and reinforcement	$F_{e,bond} = -\Delta R_{bond}$	Node reactions in AFE	$\Delta\tau_{bond}$	Stresses in ACE	Table 3.12
4	Unload and reload	a) Additional triangular deep-beam finite element of the second type for the allowance of the residual strains	$F_{e,dis,lim} = -\Delta R_{e,dis,lim}$	Node reactions in AFE	$\Delta\varepsilon_{e,dis,lim}$	Strains in AFE	Table 3.9
		b) Additional triangular deep-beam finite element of the first type for the allowance of the reload	$F_{e,rep} = -\Delta R_{e,rep}$	Node reactions in AFE	$\Delta\sigma_{crc}$	Stresses in AFE	Section 2 of this Table
5	Ultimate limit state of compression	a) Additional finite element with plastic properties s. 1 of this Table					
		b) Additional triangular deep-beam finite element of the first type for the allowance of the collapse	$F_{e,lim} = -\Delta R_{e,lim}$	Node reactions in AFE	$\Delta\sigma_{e,lim}$	Stresses in AFE	Table 3.8
6	Ultimate limit state of tension	Additional finite elements are mentioned in s. 1, 4a, 4b, 5b of this Table					

Earlier created program "Element–1" and developed now program "Element–1A" proved the fact that this procedures allow to realize nonlinear analysis of reinforced concrete structures at limit state without changing of general order of problem solving by finite element method.

Chapter 4. EXAMPLES OF NONLINEAR ANALYSIS OF REINFORCED CONCRETE STRUCTURES BY ADDITIONAL FINITE ELEMENT METHOD

The results of analysis of some reinforced concrete structures and members with use of some statements of additional finite element method are presented in this section.

First five analyses were carried out by program "Element–1" destined for nonlinear analysis of reinforced concrete structures at plane stress state. The properties of additional finite elements of the first type were used in these examples.

The choice of structures for analysis was determined by two factors: degree of transformation from stress-strain state to plane state and practical requirements.

General information of considered in analysis singularities of structures and members are listed in Table 4.1.

Table 4.1

Properties taking into account in analyzed reinforced concrete structures and members

№	**Analyzed** structure or member	Type of property taking into account			
		Plastic properties	Cracking	Bond between concrete and reinforcement	Prestressing
1	Base wall panels	+	+	–	–
2	Platform joints	+	+	–	–
3	Fibrous concrete beams	+	+	–	+
4	Beams of variable rigidity	+	+	+	–
5	Cross bar-column node	–	+	+	+

The sixth last analysis is an example of use of suggested additional finite element method in analysis of large-span reinforced concrete structures by well-known program "LYRA" realizing analysis of structures by finite element method.

4.1. Base wall panels

Necessity of analysis of base wall panels is caused by next three factors: increase of uniformly distributed load per linear meter of panel (q) due to transition from 9-floor to 10-floor construction of residential building; support of the panels on the pile heads without beam grid; decrease of concrete grade from B20 to B15 (use of concrete B15 in place of B20) [16, 97].

Analyzed variants of panels are presented in Fig. 4.1: panel 1 (УВСЦ-5) under uniformly distributed load q = 395 kN/m with seven supporting piles (Fig. 4.1a); panel 2 (УВСЦ-5) under uniformly distributed load q = 277 kN/m with five supporting piles (Fig. 4.1 b); panel 3 (УВСЦ-5-10A) under uniformly distributed load q = 387 kN/m with five supporting piles (Fig. 4.1c); panel 4 (УВСЦ-6) under uniformly distributed load q = 310 kN/m with five supporting piles (Fig. 4.1d)

As a result of calculation next characteristics of stress-strain state of the panels were obtained for each of four variants: node displacements; relative strains of finite element material; transverse, tangential and principal stresses of each finite element.

According to obtained characteristics the analysis of each variant was carried out. It includes construction of lines of principal stress, shear Q diagram and tensile N diagram for forces acting at the level of low principal working reinforcement in panels.

The main data of analysis of obtained results of calculation of all four variants of base wall panel are presented in Table 4.2. Principal stress ranges of the most unfavorable variant (panel 2 under q = 277 kN/m with five supporting piles in Fig. 4.1 b) is given in Fig. 4.2.

Table 4.2

Results of analysis of the base wall panels

<table>
<tr><th rowspan="2">№</th><th rowspan="2">Va-riant</th><th colspan="3">Principal tensile stress σ_{mt}</th><th rowspan="2">max σ_{mc}</th><th rowspan="2">Cross-sectional area of reinforcement (according tensile force N)</th><th rowspan="2">Carrying capacity of piles (according shear force Q)</th></tr>
<tr><th>$\sigma_{mt} = R_{bt}$</th><th>$R_{bt} > \sigma_{mt}$
$\sigma_{mt} > 0$</th><th>$\sigma_{mt} < 0$</th></tr>
<tr><td>1</td><td>a</td><td rowspan="4">The level of the low tensile reinforcement</td><td rowspan="2">The lesser part of the panel</td><td rowspan="2">The greater part of the panel</td><td>0,73 R_b</td><td>It is sufficient</td><td>It is sufficient</td></tr>
<tr><td>2</td><td>b</td><td>0,73 R_b</td><td>It is sufficient</td><td>It is insufficient for one pile</td></tr>
<tr><td>3</td><td>c</td><td rowspan="2">The greater part of the panel</td><td rowspan="2">The lesser part of the panel</td><td>0,88 R_b</td><td>It is sufficient</td><td>It is sufficient</td></tr>
<tr><td>4</td><td>d</td><td>0,83 R_b</td><td>It is sufficient</td><td>It is insufficient for one pile</td></tr>
</table>

This variant of panel was chosen for the test since it is met in residential buildings most often. The test confirmed the results of theoretical analysis completely. It allows to do next main conclusions:

1. The transition of the panel to support on the piles heads without grid beams changes the character of its behavior. That is why the pile heads must be in alignment to eliminate a possibility of no uniform sagging of corresponding parts of the panel as it may lead to the failure of the panel.

2. In two out of four examined variants (panel 2 under q = 277 kN/m with five supporting piles (see Fig. 4.1b) and panel 4 under q = 310 kN/m with five supporting piles (see Fig. 4.1d)) the support reaction of the panel exceed the bearing capacity (50 t). But the values of supporting reactions in both cases (53,52 t and 51,96 t) do not exceed the value provided by margin of safety 1.4 (70 t).

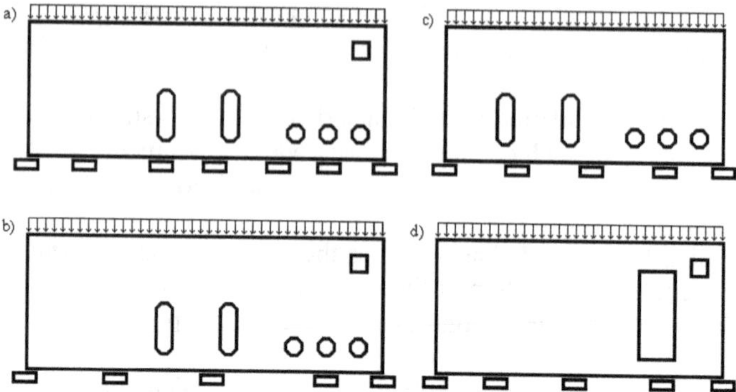

Fig.4.1. Analyzed variants of panels: a) panel 1 under $q = 395$ kN/m with seven supporting piles; b) panel 2 under $q = 277$ kN/m with five supporting piles; c) panel 3 under $q = 387$ kN/m with five supporting piles; d) panel 4 under $q = 310$ kN/m with five supporting piles

Fig. 4.2. Principal tensile and compressive stress ranges of panel 1 (УВСЦ-5) under $q = 277$ kN/m with five supporting piles (Variant b): a) principal tensile stress range (σ_{mt}); b) principal compressive stress range (σ_{mc}); 1 – $0 \leq \sigma_{mt} < R_{bt}$; 2 – $\sigma_{mt} = R_{bt}$; 3 – $\sigma_{mc} > 0{,}5R_b$; 4 – $\sigma_{mc,max} = 0{,}73R_b$

3. The base wall panels УВСЦ-5, УВСЦ-5-10А and УВСЦ-6 of good quality made of concrete grade B15 will meet the required strength, rigidity criteria and resistance to cracking under mentioned uniformly distributed loads with given number of supports. Thus the use of concrete grade B20 in place B15 will lead to cost`s cut of base wall panels due to decrease of cement and simplification of technology at casting plant.

Use of program "Element–1" for nonlinear analysis base wall panels allows to obtain a detailed phase portrait of stress-strain state of these members. Analysis of this state gives the opportunity of more rational design in comparison with procedures of SN and P [27, 28, 111].

4.2. Platform joints of large-panel building

The main problem in analysis of platform joints by program "Element–1" was the analysis of influence of different parameters of joints on their stress-strain state for more rational design (Fig. 4.3) [29].

In analysis next parameters were changed: concrete grade, sort of mortar, dimensions δ_1, δ_2, δ_3, presence of mortar in joint, eccentricity e_0. Possibility of application M100 sort of mortar in place of M200 were cleared up too.

Five variants of design of platform joints were considered:

1. Wall panels and floor panels are made of B25 concrete grade, M200 sort of mortar, width of cross joint $\delta_3 = 4{,}0$ cm, eccentricity $e_0 = 0{,}0$ cm.
2. Wall panels and floor panels are made of B25 concrete grade, M200 sort of mortar, width of bed joint $\delta_2 = \delta_3 = 4{,}0$ cm, eccentricity $e_0 = 0{,}0$ cm.
3. Wall panels and floor panels are made of B25 concrete grade, M100 sort of mortar, width of bed joint $\delta_2 = \delta_3 = 4{,}0$ cm, eccentricity $e_0 = 0{,}0$ cm.
4. Wall panels and floor panels are made of B25 concrete grade, M100 sort of mortar, width of bed joint $\delta_2 = 4{,}0$ cm, cross joint is not filled, eccentricity $e_0 = 0{,}0$ cm.
5. Wall panels and floor panels are made of B25 concrete grade, M100 sort of mortar, width of bed joint $\delta_2 = \delta_3 = 4{,}0$ cm, eccentricity $e_0 = 1{,}0$ cm.

The first three variants were directed to study stress-strain state of platform joint of good quality without eccentricity with filled joints and check of possibility of change sort of mortar in joints.

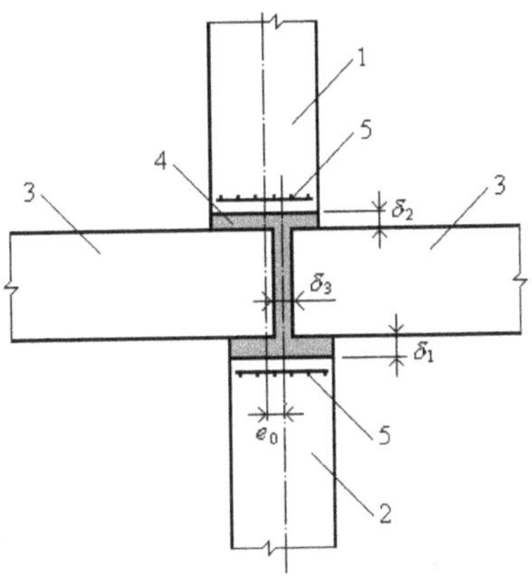

Fig. 4.3. Constructive scheme of platform joint. 1 – upper load-bearing wall panel; 2 – lower load-bearing wall panel; 3 – floor panel; 4 – mortar in joint; 5 – reinforced mesh of indirect reinforcement

The rest two variants were directed to determination of stress-strain state of platform with two defects which are met most often: absence of mortar in cross joint (variant 4) and presence of eccentricity (variant 5).

For each of five variants next characteristics of stress-strain state were obtained: node displacements, relative strains of finite element material, transverse, tangential and principal stresses of each finite element. The analysis of each variant was made according to obtained characteristics. This analysis included the construction of lines of principal stress for clearing up of dangerous areas and possible collapse of joints.

The main data of analysis of obtained results of all five variants of platform joints are presented in Table 4.3. Principal stress ranges of the most unfavorable variant 5 with eccentricity are presented in Fig. 4.4.

Table 4.3

Results of analysis of platform joint

№	Principal tensile stress σ_{mt}			max σ_{mc}	Strength for perception of required load 1100 kN/m	Mode of failure
	$\sigma_{mt} = R_{bt}$	$R_{bt} > \sigma_{mt} > 0$	$\sigma_{mt} < 0$			
1	Three FEs on the surface of the upper part of the floor panel	The greater part of the floor panel	The wall panel	$0{,}80R_b$	It is sufficient	Local crumple of the floor panel
2			The bed joint	$0{,}73R_b$		
3			The lesser part of the floor panel	$0{,}72R_b$		
4	The lower part of the wall panel and the part of the floor panel	The greater part of the wall panel and the greater part of the floor panel	The upper part of the wall panel and the greater part of the bed joint	R_b	It is insufficient	Failure of mortar in the bed joint and formation of cross cracks
5	The right floor panel and the cross joint	The greater part of the left floor panel	The greater part of the wall panel and the bed joint	R_b	It is insufficient	Failure of mortar in the bed joint

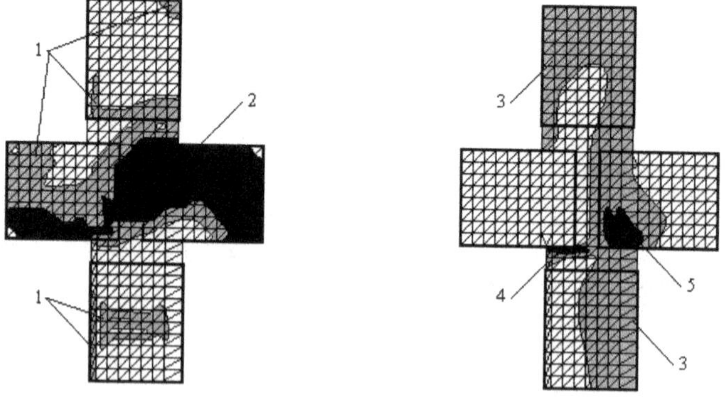

Fig. 4.4. Principal stress ranges of platform joint of variant 5:
a) principal tensile stress σ_{mt} ; b) principal compressive stress σ_{mc};
$1 - 0 \le \sigma_{mt} < R_{bt}$; $2 - \sigma_{mt} = R_{bt}$; $3 - \sigma_{mc} \ge 0{,}5R_b$; $4 - \sigma_{mc} = 0{,}74R_b$; $5 - \sigma_{mc} = R_b$

Aside from the theoretical research of platform joint the experimental test was made. This test proved the results of theoretical research completely and allowed to make next conclusions:

1. The main constructive parameters of platform joint according to their bearing capacity are next ones: quality of filling of joints by mortar especially cross joint between floor panels and presence of eccentricity in joint. Lesser influents the bearing capacity of platform joint the strength of mortar and width of joint with use of sort mortar M100 and more.

2. Platform joint without eccentricity of 16-storey residential building which is fulfilled with M100 sort of mortar with "empty" cross joint is able to carry load up to 900 kN/m. The same platform joint with filled cross joint has design load up to 1200 – 1300 kN/m. This circumstance allows to draw a conclusion that in construction of 16-storey large-panel residential buildings on the base of members of 97-th type it is possible to use in platform joints mortar M100 in place of M200 in accordance with design. It will give noticeable saving due to decrease of expenditure of concrete. It will simplify the technology of mix of mortar on the job too.

Nonlinear analysis of platform joints allowed to obtain a detailed phase portrait of its stress-strain state. Analysis of this portrait helped their more rational design in comparison with design and gave the possibility to estimate the influence of defects which are most often met during construction of large-panel residential buildings.

4.3. Fibrous concrete beams

The result of mechanical testing of prestressed beams with scheme of stresses presented in Fig. 4.5 gives the initial data of analysis. Two types of fibre reinforcement were used: wire profiled fibre and thin fibre [26, 29, 33]. Five variants of design of experimental beams are presented in Fig. 4.6:

1) БН – beam made of concrete grade B25;
2) БН-П – beam made of steel fibrous concrete grade B25 with total reinforcement by wire fibre;
3) БН-Л – beam made of steel fibrous concrete grade B25 with total reinforcement by plane fibre;
4) БН-ПС – beam of layer cross-section with internal lay of concrete grade B25 and external lays of steel fibrous concrete grade B25 with reinforcement by wire fibre;
5) БН-ЛС – beam of layer cross-section with internal lay of concrete grade B25 and external lays of steel fibrous concrete grade B25 with reinforcement by thin fibre.

The design diagram was the same in all five variants (Fig. 4.7) but the rest data were changed depending on the considered variant of beam.

The main problem of analysis of steel fibrous concrete beams was a theoretical research of their behavior for clearing up of efficiency of total substitution or partial substitution of concrete by steel fibrous concrete in cross section of beam. Besides the main problem there was an additional one connected with perfection of program "Element–1" in analysis of beams.

In this case the three main directions of study of program were chosen: the first direction was connected with development of recommendations for analysis of structures at plastic stage; the second one was connected with initial cracking; the third one was connected with possibility of allowance for prestressing of concrete by means of prestressing force P.

The sequence of theoretical analysis was next: at first the initial data with allowance for prestressing were prepared, then according to experimental values of load at which the first cracks in beams appeared the load equal to 0,98 F_{crc} (F_{crc} – theoretical load of initial cracking obtained by analysis) was gradually reached; finally the main parameters of initial cracking were determined.

Under this load next values were calculated in all five variants: node displacements, strains of finite elements, transverse tangential and principal stresses in finite elements, angles of inclinations of principal planes, location of the first crack and its direction.

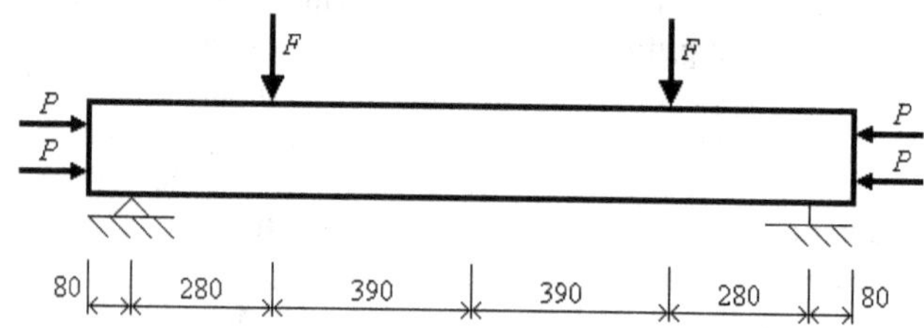

Fig 4.5. Scheme of loading of experimental beams

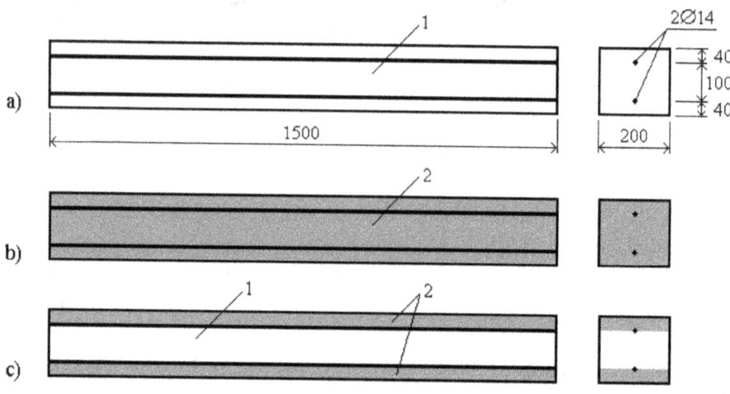

Fig. 4.6. Variants of beams: a) БН beams; b) БН-П and БН-Л beams; c) БН-ПС and БН-ЛС beams. 1 – concrete, 2 – steel fibrous concrete

Comparison of the theoretical data of analysis of five variants of steel fibrous concrete beams with the results of mechanical test of the same beams was carried out by the value of initial cracking F_{crc}, location of the first crack and its direction. The results of comparison of experimental value of load of initial cracking F^{exp}_{crc} and its theoretical value F^{theor}_{crc} for each of five variants of beams are presented in Table. 4.4. As to location of the first crack and direction of it the results of computer analysis and experimental data coincide completely: the crack appears in the middle of beam and is directed perpendicular to the longitudinal axes of beam. It is correct for all five variants of beams.

Design of beams on the basis of program "Element–1" carried out for solid structure for the allowance of the physical and mechanical properties of material of each lay allows to analyze stress-strain state. This analysis proved the efficiency of total or partial substitution in cross-section of beam of concrete by steel fibrous concrete and allowed to draw next conclusions:

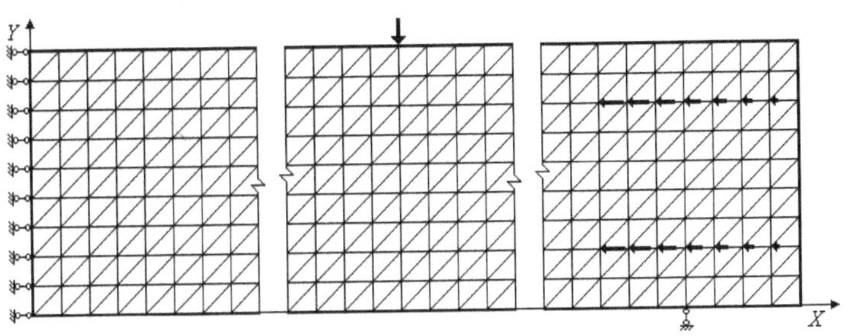

Fig. 4.7. Design diagram of the beams

Table 4.4

Experimental and theoretical values of loads of initial cracking F_{crc}, kN

№	Type of beam	F^{exp}_{crc}	F^{theor}_{crc}	Δ	|Δ| %
1	БН	33,21	32,13	+1,08	3,25
2	БН-Л	33,21	35,61	2,40	7,23
3	БН-ЛС	38,75	35,42	+3,33	8.59
4	БН-П	38,75	39,59	0,84	2,16
5	БН-ПС	44,29	38,39	+5,90	13,32

1. Introduction of fibre in concrete at all depth of beam increases its carrying capacity and simultaneously decreases straining. Therefore in equal conditions of loading and reinforcement it is possible to use steel fibre concrete structures with lesser cross-section.

2. Partial introduction of fibre in concrete at depth of beam in the places of the most stresses increases the strength as steel fibre concrete has greater strength than concrete of the same grade. In this case straining of the structure is decreased insignificantly. Therefore in equal conditions of loading and reinforcement the use of sandwich structures is more efficiently than total concrete ones.

3. In equal conditions of loading and reinforcement sandwich fibrous concrete structures are more efficient than steel fibre concrete structures as they al-

low to use their strength properties more completely. But they have more straining in comparison with solid structures.

As to estimate of use of program "Element–1" for analysis of structures and their members it should be noted that analysis with allowance for plastic properties of concrete, initial cracking and prestressing by this program allows to obtain totally reliable information of stress-strain state of analyzing structures.

4.4. Beams of variable rigidity

The main problem of analysis of four variants of beams with change of rigidity through the beam depth was the clearing up of their stress-strain state and the influence of change of this state on variations of rigid characteristics of concrete E_b, R_b, R_{bt} [29].

Together with the main problem the perfection of program "Element–1" was an additional one. The main direction of program development was connected with elaboration of recommendations for taking into account of bar reinforcement. In turn it was directly connected with requirements for reinforcement and its properties. On the other hand it was connected with reflection of singularities of bond between concrete and reinforcement. In conformity to finite element method it was connected with requirements for the properties of reinforcement finite element and connecting finite element simulating real behavior of bond between reinforcement and concrete.

Construction of the considered beams and their scheme of load are presented in Fig. 4.8. Beams were 700 mm lengthwise with cross-section 150×75 mm and one bar ∅12 A111 reinforcement.

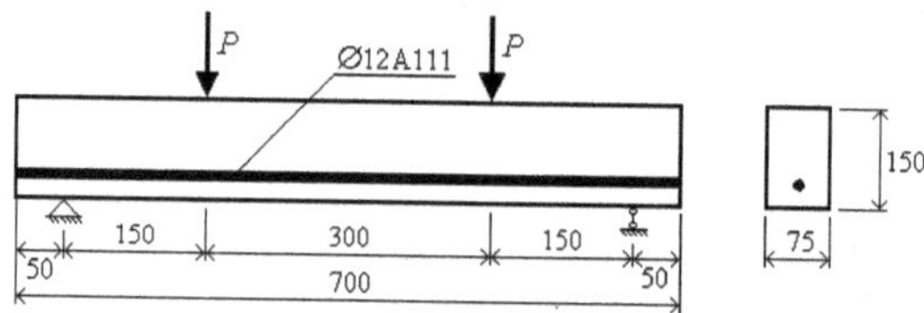

Fig. 4.8. Construction and scheme of load of beam

Four variants of variable rigidity through the beam depth were suggested for analysis:

1) Values E_b, R_b, R_{bt} are minimum at the bottom of beam at the level of reinforcement than they gradually increase up to maximum in upper third of the beam and finally partially decrease.

2) Rigid characteristics of concrete E_b, R_b, R_{bt} change inversely to the first variant; i.e. the bottom of beam is stronger than its upper part.

3) Rigid characteristics of concrete E_b, R_b, R_{bt} change in accordance with the first variant but in a sharper way. Scheme of their change is presented in Fig. 4.9.

4) Values through the beam depth are constant and in accordance with concrete of grade B25 (see Fig. 4.9).

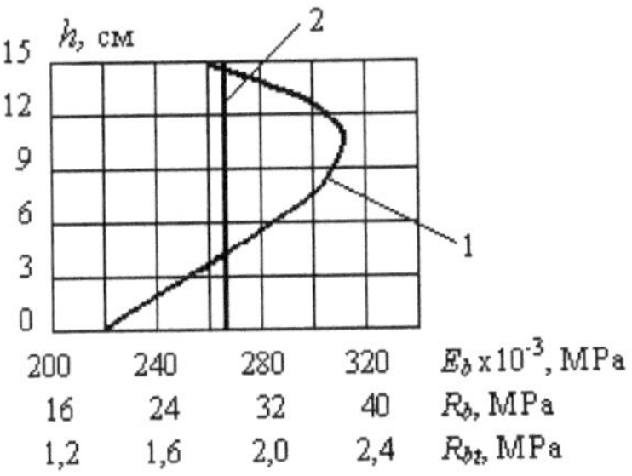

Fig. 4.9. Diagram of change of values E_b, R_b, R_{bt} through the beam depth: 1 – variant 3, 2 – in accordance of SN and P (variant 4)

Design diagram of all four variants of beams is presented in Fig. 4.10. As far as the construction and scheme of load are symmetric only the right-hand part of the beam was taken for analysis. Influence of the left-hand part was substituted by horizontal bonds along the axes of symmetry.

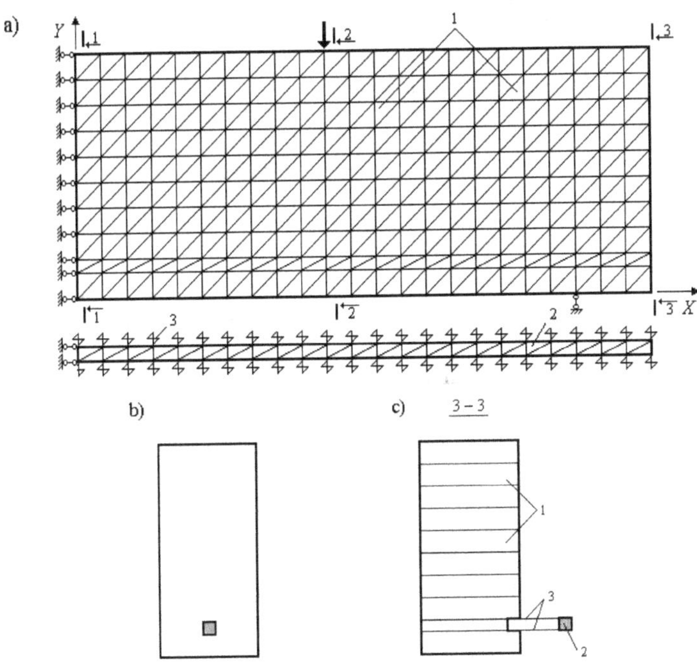

Fig. 4.10. Design diagram of beam: a) design diagram; b) equivalent cross-section; c) simulation of section (view 3–3). 1–1, 2–2 – considered sections. 1 – finite element of concrete; 2 – finite element of reinforcement; 3 – connecting finite element

Procedures of computing analysis was next: starting from experimental value of load for initial cracking in testing beam the load equals $0{,}98P_{crc}$, where P_{crc} – theoretical load of initial cracking obtained in result of computing was gradually reached. At this moment the principal tensile stresses reached the value of R_{bt} up to 2 %.

At this load for all four variants next values were determined: node displacements, strains in finite elements, transverse, tangential and principal stresses in finite elements, angle of inclination of principal planes, location of the first crack and its direction. Results of this analysis are presented in Table 4.5.

Table 4.5

Experimental and theoretical values of loads of initial cracking F_{crc}, kN

№	Type of beam	F^{exp}_{crc}	F^{theor}_{crc}	Δ	|Δ|%
1	1	5,40	5,75	–0,35	6,48
3	3	5,40	5,25	+0,15	2,81
4	4	5,40	6,50	–1,10	20,40

Table 4.6

Results of analysis of the third variant of beam. Load of cracking. Displacements.

№	Values	Load of cracking, kN		Maximum displacements depending on the character of design, 10^{-4} m					
		Elastic calculation	Plastic calculation	Elastic calculation		Plastic calculation		Taking account of the cracking	
				Along axis *Y*	Along axis *X*	Along axis *Y*	Along axis *X*	Along axis *Y*	Along axis *X*
1	P_{crc}	4,07	5,25	–5,55	1,97	–6,25	2,33	–8,50	3,13
2	%	100,0	129,0	100,0	100,0	112,6	118,3	153,2	158,9

As is obvious from obtained results the third variant of distribution of rigid characteristics through the beam depth is the nearest in values of loads of initial cracking to their experimental values and the fourth variant is the most distant one. As to location of the first crack and its direction the results are similar in all four variants: it was the normal crack in cross-section 1–1.

At the next stage of analysis two computations (elastic computation and computation with allowance of cracking) for all five variants were fulfilled to estimate the change of stress-strain state in dependence of using design apparatus.

The results obtained at this stage are presented in Tables 4.6 and 4.7. At this case it is possible to observe the redistribution of stresses due to cracking.

Table 4.7

Results of analysis of the third variant of beam. Maximum normal displacements, MPa

№	Cross-section	Values	Elastic calculation			Plastic calculation			Allowance of the cracking		
			Compressive in concrete σ_{xc}	Tensile in concrete σ_{xt}	In reinforcement σ_s	σ_{xc}	σ_{xt}	σ_s	σ_{xc}	σ_{xt}	σ_s
1	1	Stress	–2,41	1,82	13,76	–2,57	1,39	16,56	–3,56	0,00	37,66
2		%	100,0	100,0	100,0	106,6	76,41	120,3	147,7		273,7
3	2	Stress	–2,57	1,70	11,38	–2,70	1,29	13,59	–2,69	1,28	14,58
4		%	100,0	100,0	100,0	105,1	75,9	119,4	104,7	75,3	128,1

Analysis of reinforced concrete beams fulfilled by program “Element–1” for the allowance of the change of rigid characteristics of concrete through the beam cross-section depth and location of bar reinforcement allowed to carry out the analysis of stress-strain state and to make next conclusions:

1) It should be taken into account the change of rigid characteristics of concrete through the beam depth which appears during the process of production in analysis of reinforced concrete structures.

2) Among the offered variants of change of rigid characteristics through the depth of beam the third variant was the nearest to the real behavior as the deviation from test is 2,8 %.

3) In analysis in accordance with SN and P which foresees the constancy of rigid characteristics through the beam depth the deviation from theoretical data may reach 20,4 %.

4) The character of beam failures in all four variants of change of rigid characteristics of concrete is similar: failure takes place in the middle of beam transverse cross-section.

As to estimate of use of program “Element–1” in analysis of reinforced concrete and their members on the basis of carried out analysis it should be noted that analysis according this program for the allowance of the plastic properties of concrete, cracking in concrete, bar reinforcement in structure allows to obtain reliable data of stress-strain state of testing of structures.

4.5. Column-cross bar joint

Analysis of column-cross bar joint of precast reinforced-concrete framework of single-storey industrial building was fulfilled on the basis of study of creation of light-weight frame of single-storey industrial building [46].

This frame is given in Fig. 4.11. Capacity for work of the frame depends on strength of the column-cross bar joint.

Therefore the test of such joint was carried out and after its theoretical analysis was done. The length of segment of cross bar was 3,6 m. The length of segment of column was – 2,6 m.

As prestressed reinforcement strands K-7 were used. Anchor rods by length of 1,5 m were made of reinforcing steel A111.

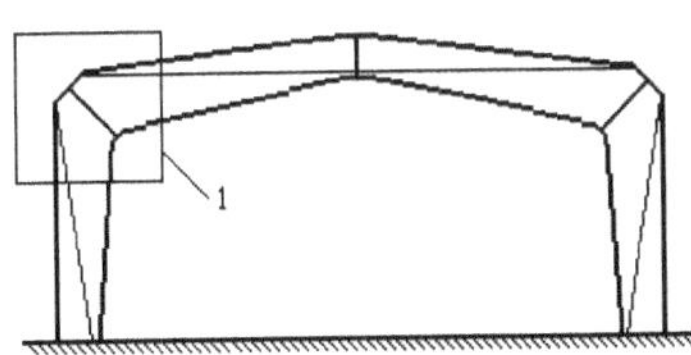

Fig. 4.11. Frame of single-storey industrial building. 1 – analyzed column-cross bar joint

In tensile zone cross bar and column were connected by tension pins with diameter 36 mm and in compressive zone they were connected by electric arc welding of inserts. Concrete grade is B30. Cross-section of cross bar changes along the length from rectangular section with dimensions 1020×300 mm to broad-flanged beam with vertical interval 820 mm. Dimensions of top flange of beam are 300×400 mm. Bottom flange has dimensions 160×120 mm. The thickness of fluted web was 60 mm.

The main problem of analysis of column-cross bar joint of frame of single-storey industrial building was the clearing out of its stress-strain state and elaboration of recommendations for its more efficient construction [40].

The check of algorithm included in program "Element–1" from point of view discrete representation of stressed bar reinforcement and unstressed one with inserts was an additional problem.

Since the column and the crossbeam are symmetric and they are fulfilled in one formwork only a segment of cross bar was calculated.

Design diagram is presented in Fig. 4.12a. The types of cross bar are shown in Fig. 4.12b. Due to shortage of financing it was necessary to restrict analysis of elastic problem only. Such realization does not allow to take into account plastic properties of concrete. But the preliminary estimate of stress-strain state of structure according to the results is possible.

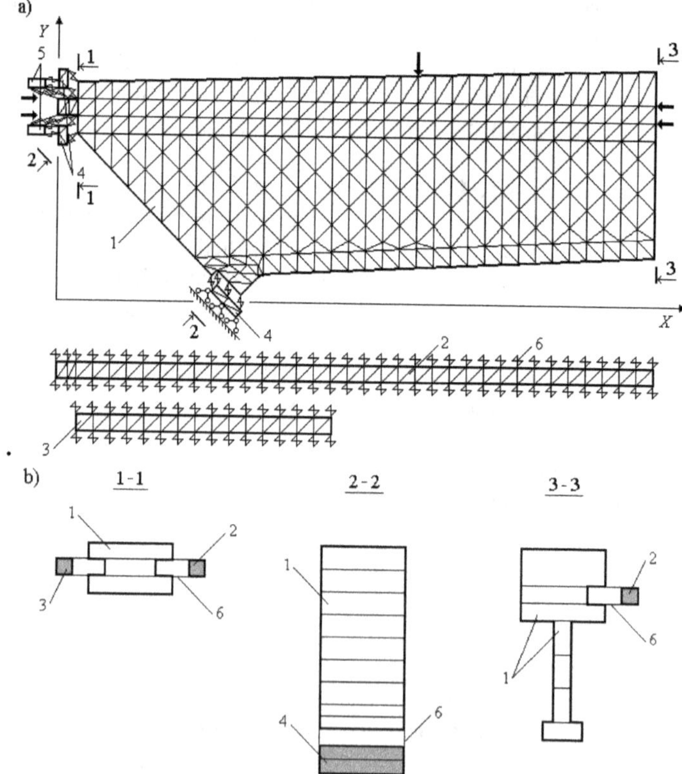

Fig. 4.12. Design diagram of column-cross bar joint: a) design diagram; b) types of transverse sections. 1 – finite element of concrete; 2 – finite element of stressed reinforcement; 3 – finite element of unstressed reinforcement; 4 – finite element of inserts; 5 – finite element of anchorage device; 6 – connecting finite element

Elastic analysis of column-cross bar joint was carried on by two stages for more precision. The aim of the first stage was the determination of initial cracking load (which had the value of 38 t) and behavior of connecting finite elements. The second stage was carried out under this load but rigidities of longitudinal bonds of connecting finite elements were determined without allowance for mutual displacement of concrete and reinforcement. Rigidities of transverse bonds were introduced depending on their behavior in compression and tension.

Boundary node displacements, scheme of deformation and prognostic cracking (Fig. 4.13a), values of transverse, tangential and principal stresses in separate finite elements were obtained.

The cross section 1–1 which stress sheet is shown in Fig. 4.13b turned out to be the most dangerous one. Besides this cross-section two other were considered: cross-section at the place of break of anchor rods and cross-section under the acting force.

Comparison of results of analysis with test data exhibited that the pictures of real cracking and prognostic cracking coincided. But real displacements are more than theoretic ones: the difference in separate cases reached 24,6 %. It is explained by influence of plastic properties of concrete.

Analysis of column-cross bar joint of frame of single-storey industrial building fulfilled by program "Element–1" allows to analyze it stress-strain state and to make next conclusions:

1. Failure of column-cross bar joint takes place at cross-section 1–1 which is situated over the insert at the point of support. Therefore it is necessary to stregnthen the cross-section to increase the bearing capacity of joint.

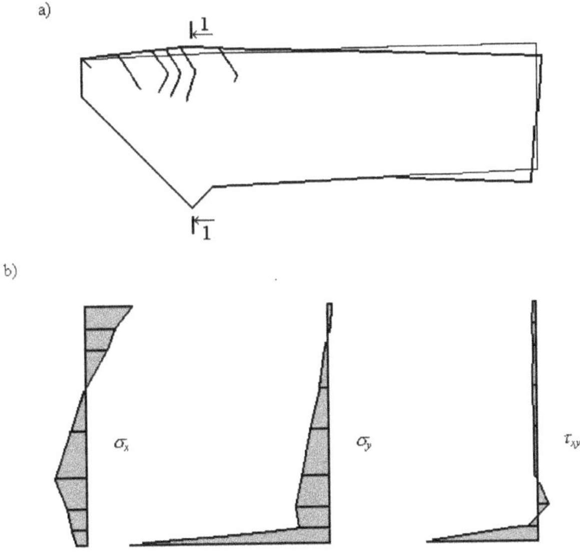

Fig. 4.13. Results of analysis: a) scheme of deformation and prognostic cracking; b) stress sheet in dangerous cross-section 1–1

2. Analysis of stresses in cross bar of joint exhibits that it is possible to take for test the segment of lesser length but not less than 2,4 m.

3. Distribution of tangential stresses along the length of anchor rod shows that its length may be lesser but not less than 1,2 m.

4. To increase the bearing capacity of column-cross bar joint it is necessary to study in details the behavior of anchor device for transmission the stress from the reinforcing steel to the concrete by means of inserts.

As to estimate of use of program "Element–1" in analysis of reinforced concrete structures that on the basis of fulfilled analysis we may say next:

1. Program may be used in analysis of structures with irregular shape and irregular net of design diagram. It allows to study in detail the behavior of separate places which are interesting from the point of view of engineer.

2. Program may be used in analysis of structures with stressed reinforcement, unstressed reinforcement and in the presence of inserts.

The fulfilled analysis of reinforced concrete structures and their members shows:

1. Allowance for the specific singularities of behavior of the reinforced concrete structures (plastic properties of concrete, cracking, nonlinearity of bond between concrete and reinforcement, prestressing and bar reinforcement) in analysis by finite element method gives the opportunity to obtain more real information of stress-strain state of these structure that leads to rational design.

2. Comparison of theoretical and testing results exhibits that the developed program "Element–1" describes the stress-strain state sufficiently precisely since the results coincide in quality and quantity.

3. Algorithms for the allowance of the nonlinear properties on the basis of use of additional loads proved their efficiency.

4.6. Large-span shell of Chelyabinsk Shopping Centre

Analysis of the shell of Chelyabinsk Shopping Centre was fulfilled to prove the perspective of created procedures for a thin shell. This analysis is a part of many years research under the guidance of professor Maximov Y.V. at department of building structures of South Ural State University. It is destined for estimate of influence of daily temperature difference on behavior of large-span reinforced concrete shells.

Chelyabinsk Shopping Centre was constructed in 1975 [13, 14, 60, 61, 75, 99, 103–107]. It has the square hall with an area of more than 10000 square meters. Building represents a part of sphere cut off from four quarters by vertical planes. The shell is formed by translating a plane generetrix (arc of 264 m dia) along parallel guides (arcs of the same dia with a chord of 102 m). The rise of arc is equal to 10,2 m, i.e. the rise of the shell in the centre equals 20,4 m (Fig. 4.14a).

The main load bearing part of the shell is prestressed concrete contour, which hinges the pendulum columns with spacing equals 6 m. The shell is attached to a slide on monolithic bases in every corner. The reinforced concrete shell represents the complicated cast-in-place and precast structure parted into curved rhombic quadrangles. The horizontal projection of each curved rhombic quadrangles is the square with 12×12 m in dimensions. The prestressed plates with 3×12 m in dimensions are built in these parts. The shell is jointed with contour.

The slides guarantee a free displacement of corner points under vertical load along the lines parallel to corresponding diagonal of the building.

Two opposite sides of the square of Chelyabinsk Shopping Centre are oriented meridionally, i.e. one of them looks South, another one looks North (Fig. 4.14б).

The shell was calculated for uniformly distributed load of 650 kg/m^2 (dead weight, suspended absorbing ceiling, warming- keeping jacket, ruffling and snow).

In addition the reconstruction of the roof was made in 1989. As a result of it the load due to dead weight of the shell was increased by 4.5 %.

An analogous shell of Minsk Central Market-place in dimensions 103×103 m was constructed in 1979. Unlike to the shell in Chelyabinsk this shell is simply supported on steel plates at the corners. Also this building has another orientation. One diagonal is oriented meridionally from North to South; another one is oriented latitudinally from East to West (Fig. 4.14c).

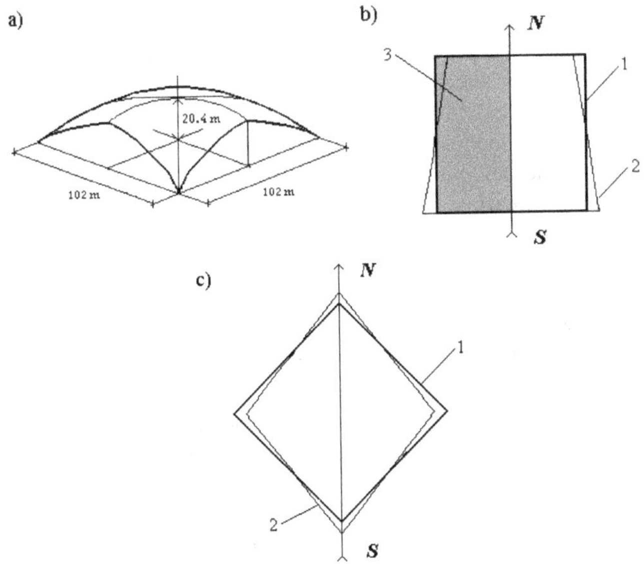

Fig. 4.14. Orientation of shells: a) general view; b) shell in Chelyabinsk; c) shell in Minsk; 1 – design position; 2 – real position; 3 – part of the shell for calculation

Since the building of Chelyabinsk Shopping Centre is unique the operation conditions of the construction are under the regular control. Some unfavorable factors were discovered during these observations, such as cracks in joints and cracks in adjacent contour zone. The cracking was observed in this zone from the South in spring and autumn probably due to irregular heating of the shell by the sun.

Simultaneously as a result of control of Minsk Central Market-place it turned out that it changed the form of the plane of reference, i.e. the diagonal oriented meridionally was elongated, on the contrary the latitudinal oriented diagonal became shorter (Fig. 4.14c). It was supposed that the irregular heating of the shell by the sun was at bottom of it. The measurement shows that the change of the length of the contour member is equal to 1,2 mm during the change of 1 degree centigrade in temperature.

It is supposed that in Chelyabinsk too under the temperature action especially in spring and autumn the south side of the shell became longer and the north side became shorter thus the additional load of the shell was created. This supposition was proved by the fact that during prestressing of members of contour the temperature differences on the surface of metal details of south side with respect to the north side was from 50°C to 60°C. At the same time it is necessary to check this supposition in theory and in practice by means of investigation of the shell.

Since now it is impossible to carry out the detailed investigation of state of the shell due to absence of funds we must restrict ourselves to some theoretical analysis, which does not demand considerable expenses and labor expenditure. Analysis by finite element method was one way of this theoretical analysis [30–32, 59, 117, 118].

In order to check suppositions appeared from estimate of results of regular control an analysis of the shell was executed with the help of application package "LIRA" ("MIRAGE") realizing the analysis of reinforced concrete structures by means of finite element method.

The analysis had next purposes:

1. estimation of bearing capacity of the shell in connection with increased load of the roof after its reconstruction;

2. estimation of the action of temperature on bearing capacity of the shell;

3. perfection of procedures of analysis of the shell under the action of temperature on the basis of additional load which was formed on the properties of the corresponding additional finite elements, limit states of the main finite elements and an accepted failure model of the structure.

4. estimation of all potentialities of the application package "LIRA" destined for linear analysis of reinforced concrete structures at limit state by finite element method.

These problems were solved simultaneously in accordance with stated below sequence. Since the program "LIRA" was used in analysis of the shell the design diagram (see Fig. 4.15) was prepared in accordance with principles and library of finite elements of this program. Two types of finite elements were used for a simulation of the shell of the Shopping Centre: type № 5 (the bar of general attitude) and type № 44 (the plane element of the shell of general attitude). The first type was used for simulation of the elements of index contour, columns and cross-beams. The second one was used for simulation of the plates of roof cover.

Since the action of temperature is symmetric about the axis oriented from North to South only a half of the structure was examined. The analysis of the action of temperature was made under the following initial conditions: the temperature difference in meridian direction (from North to South) was 40°C; the temperature inside the building was +20°C. All potentialities of computer were used in analysis of action of temperature.

For example, two procedures of representation of the temperature action were taken into consideration: for bar finite elements in the form of tensile and compressive forces at the ends of the shell. The second variant was based on the properties of additional finite elements and limit states of the main finite elements. This variant tuned out more real and it will be considered later. Prestressing was simulated by means of compressive load along the shell contour. Singularities of representation of heat load and prestressing will be explained further because at first some questions connected with the realization of limit state analysis should be considered.

For beginning of the limit state analysis of structure it is necessary to determine the type of limit state which corresponds to this type of action on the structure. It allows to choose a failure model in the considered case [38, 40].

As the results of regular control shows the cracks appeared in concrete of member on south side of index contour.

It means that the prestressing is insufficient for overcoming of unfavorable influence of temperature. Thus the limit state of the structure is the state before failure of concrete at south side of index contour due to tensile strains of increased temperature action.

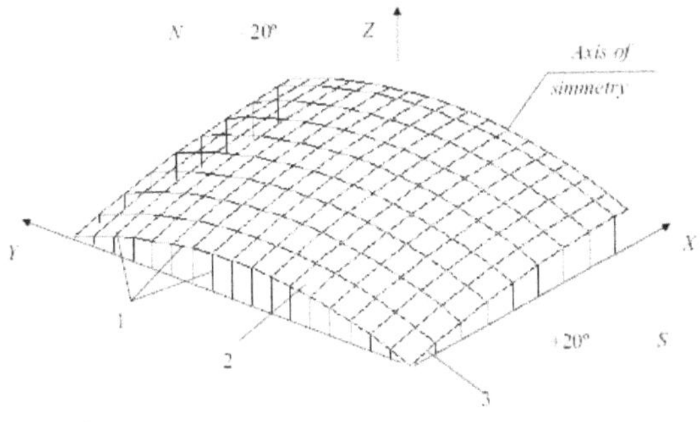

Fig. 4.15. Design diagram: 1 – finite element of type 5; 2 – finite element of type 44; 3 – finite element of index contour under the most unfavorable combination of loads

Therefore a failure model of the shell in this case is the initial design diagram without one bar finite element simulating the south side of index contour. Such approach means that ultimate limit state of the structure is determined by ultimate limit state of the only bar finite element. This element is the nearest one to the south point of support of the shell (Fig. 4.15). There is the most unfavorable combination of loads in this bar finite element: pitch component of load, prestressing and action of temperature. It allows to simplify essentially the problem if the properties of additional finite elements and ultimate limit states of the main finite element are used. The ideal failure model of the shell was proposed in the previous section. It was the initial design diagram of structure where the finite element nearest to the south point of support is excluded. It means that ultimate limit state of the structure is determined by the ultimate limit state of this finite element. For solving of the problem it is necessary to determine the type of this ultimate limit state and the conditions of its appearance.

The finite element bar of general attitude (s. In.1.5) was used for simulation of index contour. System of strains of this element and its corresponding limit states were considered in s. 1.2.4.

Since the results of control showed cracking in the nearest to the corner point of the south side of index contour (Fig. 4.15) that it is possible to choose the limit state due to tension by longitudinal force (s. 1.2.4 and Table 1.1).

This limit state of the given element determines the limit state of the whole shell in solving of the formulated problem. Such approach simplifies the calculation essentially.

As long as program "LIRA" was used for solving of the formulated problems of analysis of the shell then before the beginning of the analysis the question of its behavior due to outer action on it arises.

In this case these singularities were next:

1) increasing of the dead weight of the shell due to its reconstruction;
2) prestressing of the members of the index contour;
3) temperature action due to difference of temperature between the north side of the shell and the south one.

The first problem was solved by the simple increasing of the design uniformly distributed load. The solving of two others required more complicated approach.

At first we turn our attention to allowance for presstressing. The definition of real portrait of stress-strain state of prestressed structure is impossible without estimate of influence of prestressing on it since prestressing is a creation of the opposite stresses before the moment of application of the main stress.

Thus the resulting stress-strain state of prestressed structure is the result of two main actions: at first of prestressing of an unloaded structure and then of service load.

This circumstance creates definite difficulties of analysis of prestressed structures by any methods, finite element method including. The main problem is a simulating of prestressing.

The way of simulating of prestressing by application of corresponding forces of prestressing is well-known. This way is based on the accepted by SN and P approach to calculation of prestressed structures.

This way gives a good result not only in analysis of bending beams but also in analysis of space frame constructions as in this case the general approach included in application of outer load after prestressing remains unchanged. This approach was used in analysis of the shell of Shopping Centre.

In this structure the prestressed strands situated in four corners and were used for prestressing. It created a system of prestressing in two mutually perpendicular directions.

Prestressing force of index contour P along the side of the support's square is an irregular distributed load in accordance with tensile stress sheet and strand reinforcement diagram. The number of strands was changed according to the value of tensile stresses, therefore the value of prestressing force was applied to nodes of design diagram step-by-step, i.e. $P = P_1 + \ldots + P_i \ldots + P_n$, where $n = 8$ – the number of nodes of a half of one side of index contour, in which the corresponding parts of prestressing force P were applied. Since the conditions restricted the displacements are applied in supporting points of corners, then the part of prestressing force P_1 for these nodes ought to be applied in definite distance from nodes.

Described above procedures for the allowance of the prestressing as an outer additional load meet next requirements completely:

1) accepted in SN and P approach to consideration of prestressing as an outer action of force;

2) logic of analysis of structures by finite element method;

3) possibilities of the program "LIRA" for realization of taking account of prestressing in computer calculation.

Now we turn our attention to determination of additional load connected with difference of temperature between the north side of shell and the south one. This process is based on combination of properties of accepted design diagram of structure with formation of additional load using singularities of additional finite element of bar of general attitude.

Since the shell of Shopping Centre was divided into 17 rows of the plane finite elements of the shell of general attitude (type № 44) then according to difference of temperature 18 boundary nodes for application of the vectors $F_{t,i}$ simulating temperature action was considered.

Thus the general vector of additional load for the allowance of the difference of temperature may be determined as [32]:

$$F_t = \sum_{i=1}^{n} F_{t,i} , \tag{4.1}$$

where $F_{t,i}$ = vector of additional load for each boundary node of design diagram; n = 18 boundary nodes.

For definition of value of such vector $F_{t,i}$ we may use the properties of additional finite element of space bar for the allowance of the limit state of the corresponding main finite element at ultimate limit state of tension.

Choice of ultimate limit state of the main finite element is determined by the accepted failure model of the shell due to tension of south side of index contour. The additional finite element of the second type changing the strain state of the main finite element may be used to make allowance for its ultimate limit state and for formation of additional load with allowance for action of temperature.

Therefore the vector of additional load may be determined by the formula:

$$F_{t,i} = E_b A_i \Delta l_i / l , \tag{4.2}$$

where E_b = modulus of elasticity of concrete; A_i = area of cross-section; $\Delta l_i / l$ = relative elongation in changing on the value Δt_i calculated as:

$$\Delta l_i / l = \alpha \Delta t_i \,, \tag{4.3}$$

where α = linear expansion coefficient of concrete;

l = length of the side of index contour.

As to the value of difference of temperature Δt_i it is determined from condition that the temperature of the south side of index contour is +20°C and the north side of index contour has the temperature – 20°C, i.e. difference of temperature is 40°C.

Thus in this analysis the general additional load of the shell consists of two loads: the additional load of prestressing P and the additional load F_t for the allowance of the temperature difference in direction from North to South.

Introduction of additional load simplifies essentially the solving of problem as it allows to use program "LIRA" realizing linear analysis of structure by finite element method.

Analysis of results shows next:

1) The stress in the shell elements increases due to the reconstruction of the shell, but depending on the location it turns out from 25% to 78% less than limit characteristics of reinforcement.

2) The increase of the tensile forces in the elements of index contour due to irregular heating of the shell by the sun and the temperature difference from North to South up to 17°C does not exceed the limit design values.

3) Allowance for the action of temperature the stresses in the plates of the roof turned out from 25 % to 41 % less than the rated values.

4) The most dangerous place of the shell is a zone near the south side of the prestressed index contour.

5) It is necessary to take into account the influence of daily temperature difference on behavior of long-span shells in their design.

As the control showed there are cracks in concrete in adjacent contour zone of the south side of index contour, i.e. in this place the prestressing force is insufficient for amortization of tensile stresses appearing at heating of the shell by the sun.

It coincides completely with the results of theoretical analysis of temperature action fulfilled on the basis of accepted ideal failure model of structure and additional temperature load determined on the basis of properties of the additional finite element of the bar of general attitude with allowance for limit state in tension of the corresponding main finite element.

The results of fulfilled theoretical analysis of the shell of Shopping Centre by program "LIRA" allow to make next conclusions:

1. Qualitative coincidence of theoretical data with result of control showed that the accepted procedure describes sufficiently accurate the ultimate limit state of the considered structure.

2. The proposed procedures based on the use of additional loads calculated by the additional finite elements allow to realize limit state analysis of space structure.

3. This procedure opens the opportunity to use the earlier developed programs of linear analysis for realization of limit state analysis of the structures.

4. Representation of ideal failure model of space structures opens wide possibility for limit state analysis of these structures.

Unfortunately the further study of behavior of the shell of Chelyabinsk Shopping Centre was ceased due to absence of finances.

Fulfillment of analysis of reinforced concrete structures and their members by program “Element–1” allows to make next conclusions:

1. The proposed procedure based on the use of the properties of additional finite elements allows to realize the analysis of plane and space reinforced concrete structures at limit states.

2. It is possible to realize limit state analysis of reinforced concrete structures by two ways:

a) by means of programs formed according to the algorithms based on this procedure and the earlier developed linear programs of analysis of structures by finite element method;

b) by means of the existing programs with use of additional loads calculated on the basis of properties of the additional finite elements.

3. The representation of ideal failure model of plane and space structures helps to realize limit state analysis of these structures by finite element method.

CONCLUSIONS

The result of fulfilled research allows to make next conclusions:

1. Elaborated method of analysis of reinforced concrete structures at limit state by means of additional finite elements (AFEM) contains the next main statements:

a) AFEM is a variant of nonlinear finite element method representing combination of the three well-known methods: finite element method, method of additional loads and limit state method;

b) AFEM is based on use of two well-known concepts: the ultimate limit state of structure and the ultimate limit state of finite element;

c) this method supposes the information of behavior of each finite element as each of its characteristic ultimate limit state is reaching;

d) process of change of nonlinear properties of the separate finite element at reaching step-by-step of ultimate limit state is simulated by special developed additional finite elements allowing to change the properties of the main finite element;

e) proposed additional finite elements allow to form the vectors of additional loads independently of the character of the observed nonlinear properties of the main finite elements;

f) additional loads may be formed by the three ways on the basis of: additional finite elements of general type changing the stiffness matrices of the main finite element; additional finite elements of the first type changing the stress state of the main finite elements; additional finite element of the second type changing the strain state of the main finite elements;

g) it is proposed to use an ideal failure model which represents design diagram of the structure at the moment of ultimate equilibrium for description of ultimate limit state of the whole structure;

h) use of ideal failure model of structure opens wide opportunities for realization of limit state analysis by linear variant of finite element method and nonlinear one;

i) in any type of analysis of structure by finite element method the ideal failure model may be formed by analysis of behavior of the structure under step-by-step increasing load or by preliminary representation based on the acting SN and P, the analysis of behavior of analogous structures or the results of their investigation;

j) in use of additional finite elements the ideal failure model is formed by change of the properties of the main finite elements entering the initial design diagram of the considered structure;

k) design diagram consists of two ones: the design diagram formed from the main finite elements with linear properties and the design diagram formed from additional finite elements with nonlinear properties corresponding to the given stage of limit state;

l) design diagram from additional finite elements changes the initial design diagram consisting of linear finite elements thus that it may correspond to the reached at this moment stage of the limit state of structure; at ultimate limit state it turns the initial design diagram into the ideal failure model of structure;

m) proposed procedure is sufficiently simple and may be the base for algorithms and programs realizing limit state analysis of structures.

2. Algorithms and programs destined for limit state analysis of reinforced concrete structures are developed in accordance with the proposed procedures and they are characterized by next aspects:

a) earlier created algorithms and programs for linear and nonlinear analysis of structures by finite element method may be assumed as their basis;

b) they may be realized as independent programs destined for limit state analysis or as auxiliary programs complementary for existed ones;

c) in realization of these programs as auxiliary ones it is necessary to foresee two their components: for formation of additional load in representation of initial data and for calculation of correcting values of obtained results;

d) these programs ought to have special blocks destined for analysis of properties of the separate finite elements depending on the reached stage of limit state typical for the given type of failure;

e) in analysis of reinforced concrete structures at plane stress state the programs considered the behavior of each finite element depending on the stage typical for its limit state: plastic behavior and failure in compression; plastic behavior before cracking, partial unload after cracking, behavior with crack, failure in tension;

f) programs ought to have special blocks destined for analysis of the whole structure depending on the stage of its limit state typical for the given type of failure, i.e. blocks of comparison of it`s behavior with an ideal failure model;

h) in analysis of bending beams at plane stress state the programs ought to foresee a possibility of use of the two models of failure accepted in SN and P: with transverse crack and oblique one;

i) in realization of limit state analysis of the structure with use of additional load these programs essentially reduce calculations at the expense of operation of direct execution of Gauss method in solving of algebraic system of linear equations;

j) programs are the flexible changeable instrument for calculation which allows a rapid interference into any block for perfection of earlier represented algorithms;

k) programs allow to use different theories and procedures at separate stages of behavior of structure at reaching its limit state;

l) programs may be used for analysis of structure under definite load and for computer research of its behavior from the beginning of load to the moment of ultimate limit state.

3. Fulfilled analysis of beams by program "Element–1" and the shell of Chelyabinsk Shopping Centre by means of program "LIRA" with use of additional loads allows to make next conclusions:

a) taking account of behavior of the reinforced concrete structures depending on the reached stage of limit state makes it possible to obtain more real information of their stress-strain state;

b) comparison of theoretical and testing results showed that the developed procedure sufficiently correct describes stress-strain state of the structure as the coincidence in quality and quantity was observed;

c) use of procedure based on use of properties of the additional finite elements and the ideal failure models open wide possibility in realization of limit state analysis of reinforced concrete structures.

REFERENCES

1. Агапов В.П. Метод конечных элементов в статике, динамике и устойчивости пространственных тонкостенных подкрепленных конструкций. – М.: АСВ, 2004. –248 с.

2. Алмазов В.О. Евронормы и СНиП'ы: Сборник материалов 2-й Всероссийской конференции по проблемам бетона и железобетона. – М.: 2005. – Т. 2. – С. 17 – 27.

3. Алямовский А.А. SolidWorks/COSMOSWorks. Инженерный анализ методом конечных элементов. – М.: ДМК Пресс, 2004. – 432 с.

4. Аргирис Дж. Современные достижения в методах расчета конструкций с применением матриц. – М.: Изд-во лит. по строительству, 1968. – 241 с.

5. Багаев Б.М., Шайдуров В.В. Сеточные методы решения задач с пограничным слоем. – Новосибирск: Наука. Сиб. предприятие РАН, 1998. – Ч. 1. – 199 с.

6. Байков В.Н., Сигалов Э.Е. Железобетонные конструкции. Общий курс. – М.: Стройиздат, 1991. – 767 с.

7. Берг О.Я. Некоторые вопросы теории деформаций и прочности бетона // Строительство и архитектура. – 1967. – № 10. – С. 41–56.

8. Биргер И.А. Круглые пластинки и оболочки вращения. – М.: Оборонгиз, 1961. – 364 с.

9. Биргер И.А. Общие алгоритмы решения задач теории упругости, пластичности и ползучести: Успехи механики деформируемых сред. – М.: Наука, 1975. – С. 61–73.

10. Биргер И.А. Стержни, пластинки, оболочки.– М.: Наука, 1992. – 362 с.

11. Бондаренко В.М., Суворкин Д.Г. Железобетонные и каменные конструкции.– М.: Высшая школа, 1987. – 364 с.

12. Бондаренко В.М., Судницын А.И., Назаренко И.Г. Расчет железобетонных конструкций и каменных конструкций. – М.: Высшая школа, 1988. – 303 с.

13. Важенин Ю.П., Черкасов В.А. Монтаж железобетонной оболочки размерами 102х102 м // Реферативная информация о передовом опыте. – 1975. –Вып. 7(76). – С. 68.

14. Васильев Н.Ф., Максимов Ю.В., Марков В.А. Оборудование для предварительного обжатия крупноразмерной оболочки // Транспортное строительство. – 1980. – №1. – С. 24–26.

15. Власов В.З. Избранные труды. – М: Изд-во АН СССР, 1962. – Т. 1. – 580 с.

16. Внутренние стеновые панели и стенки лоджий. Альбом Ч. 10, р.10.2-1.–Челябинск, 1980. – 96 с.

17. Гвоздев А.А. Расчет несущей способности конструкций по методу предельного равновесия: Сущность метода и его обоснование. – М: Госстройиздат, 1949. – 280 с.

18. Гениев Г.А., Киссюк В.Н., Тюпин Г.А. Теория пластичности бетона и железобетона. – М: Стройиздат, 1974. – 316 с.

19. Голованов А.И., Закиров Р.Ф., Паймушин В.Н. и др. Расчет напряженно-деформированного и предельного состояния железобетонных кон-

струкций. – Труды международной конференции «Численные и аналитические методы расчета конструкций». – Самара, 1998. – С. 67–71.

20. Голышев А.Б., Бачинский В.Я., Полищук В.П. и др. Проектирование железобетонных конструкций. Справочное пособие. – Киев: Будивэльник, 1990. – 544 с.

21. Голышев А.Б., Полищук В.П., Руденко И.В. Расчет железобетонных стержневых систем с учетом фактора времени. – Киев: Будивэльник, 1984. – 128 с.

22. Гольштейн Р.В., Салганик Р.Л. Хрупкое разрушение тел с произвольными трещинами: Успехи механики деформируемых сред. – М.: Наука. 1975. – С. 156 –171.

23. Городецкий А.С., Евзеров И.Д., Стрелец-Стрелецкий Е.Б. и др. Метод конечных элементов: теория и численная реализация: Программный комплекс «Лира – Windows». – Киев: Факт, 1997. – 138 с.

24. Дарков А.В., Шапошников Н.Н. Строительная механика. – М.: Высшая школа, 1986. – 608 с.

25. Ермакова А.В. Метод дополнительных конечных элементов для расчета железобетонных конструкций по предельным состояниям: Монография. М., Издательство строительных ВУЗов, Челябинск, Издательство ЮУрГУ, 2007 – 128 с.

26. Ермакова А.В., Зива А.Г., Соловьев Б.В., Боксбергер Э.К. Влияние фибрового армирования на деформативность и трещиностойкость изгибаемого элемента с учетом слоистости конструкции: Фибробетон: свойства, технология, конструкции. Тезисы докладов. 26–27 апреля 1988. – Рига: ЛатНИИстроительства, 1988. – С. 41–43.

27. Ермакова А.В., Карякин А.А. Исследование работы стеновых цокольных панелей жилых зданий на ЭВМ // Совершенствование железобетонных конструкций для промышленного и гражданского строительства и технология их изготовления на Среднем Урале: Тезисы докладов. 5–6 мая 1988. – Свердловск: НТО, 1988. – С. 13–14.

28. Ермакова А.В. Расчет железобетонных конструкций, работающих в условиях плоского напряженного состояния, с помощью программы «ЭЛЕМЕНТ–1»: Исследования по бетону и железобетону. – Челябинск: ЧПИ, 1989. – С. 15–18.

29. Ермакова А.В. Расчет железобетонных конструкций с помощью программы «ЭЛЕМЕНТ–1»: Исследования по строительным материалам, конструкциям и механике. – Челябинск: ЧГТУ, 1991. – С. 63–68.

30. Ермакова А.В., Максимов Ю.В. Расчет оболочки покрытия торгового центра в г. Челябинске на температурные воздействия: Научные труды Общества железобетонщиков Сибири и Урала. – Новосибирск: СГАПС, 1994. – С. 44 – 47.

31. Ермакова А.В., Максимов Ю.В. О расчете оболочки покрытия торгового центра в Челябинске на динамические воздействия: Научные труды Общества железобетонщиков Сибири и Урала. – Новосибирск: СГАПС, 1995. – С. 63 – 66.

32. Ермакова А.В., Максимов Ю.В., Оатул А.А. Анализ работы покры-

тия торгового центра г. Челябинска с учетом действия температуры: Исследования по строительным материалам, конструкциям и механике. – Челябинск: Изд-во ЧГТУ, 1996. – С. 81 – 85.

33. Ермакова А.В., Максимов Ю.В. Учет предварительного напряжения при расчете железобетонных конструкций методом конечных элементов: Бетон на рубеже третьего тысячелетия: Сб. мат-в 1-й Всероссийской конференции по проблемам бетона и железобетона. – М., 2001. – Кн. 2. – С. 968–973.

34. Ермакова А.В. Учет нелинейных свойств железобетона с помощью дополнительных нагрузок при расчете методом конечных элементов: Бетон на рубеже третьего тысячелетия: Сб. мат-в 1-й Всероссийской конференции по проблемам бетона и железобетона. – М., 2001. – Кн. 2. – С. 974–979.

35. Ермакова А.В. Треугольный конечный элемент балки-стенки с условной трещиной // Вестник ЮУрГУ. Серия «Строительство и архитектура». – 2003. – № 7(23). – Вып.2. – С. 37 – 40.

36. Ермакова А.В. Расчет конструкций по предельным состояниям с использованием метода конечных элементов: Пространственные конструкции зданий и сооружений. Вып. 9 / МОО «Пространственные конструкции». – М.: ООО «Девятка Принт», 2004. – С. 16 – 25.

37. Ермакова А.В. Дополнительные нагрузки для расчета конструкций по предельным состояниям методом конечных элементов // Вестник УГТУ-УПИ №11(41). «Строительство и образование»: Сборник научных трудов. – Екатеринбург: ГОУ ВПО «УГТУ-УПИ», 2004. – С. 100 – 102.

38. Ермакова А.В., Максимов Ю.В. Модели разрушения железобетонных конструкций при расчете методом конечных элементов: Тезисы докладов научной сессии «Расчеты и проектирование пространственных конструкций с учетом физической и геометрической нелинейности» МОО «Пространственные конструкции» 15 декабря 2004. – М.: МОО «Пространственные конструкции», 2004. – С. 28–29.

39. Ермакова А.В. Расчет железобетонных конструкций по предельным состояниям методом дополнительных конечных элементов: Сборник материалов 2-й Всероссийской конференции по проблемам бетона и железобетона. – М.: 2005. – Т. 2. – С. 386 – 391.

40. Ермакова А.В.и Максимов Ю.В. Идеальные модели разрушенияядляярасчетаяжелезобетонныхяконструкцийяпояпредельнымясос2ояниямя ме2одомя конечныэ я элемен2ов:я Сборникя ма2ериаловя 2-йя Всероссийскойяконференциияпояпроблемамябе2онаяияжелезобе2онаэя–я Мэ:я2005эя–яТэя2эя–яСэя392я–я397эя

41эяЕрмаковаяАэВэяТеоре2ическиеяосновыяме2одаядоСолни2ельфныэяконечныэяэлемен2овядляярасче2аяжелезобе2онныэяконс2рЗкцийяСояСредельнымясос2ояниямэяСбэяс2а2ейэяВыСэя10эяМООя«Прос2ранфс2венныеяконс2рЗкции»;яСодяредэяШЗгаеваяВэВэя–яМэ:я2006иясэя30я–я41э

42. Забегаев А.В. К построению общей модели деформирования бетона // Бетон и железобетон. – 1994. – № 6. – С. 23 – 26.

43. Зенкевич О.К. Метод конечных элементов в технике.– М.: Мир, 1975.– 541с.

44. Зенкевич О., Чанг И. Метод конечных элементов в теории сооружений и механике сплошных сред. – М.: Недра, 1974. – 238 с.

45. Зенкевич О., Морган К. Конечные элементы и аппроксимация. – М.: Мир, 1986. – 318 с.

46. Ивашенко Ю.А., Лобанов А.Д. Исследование процесса разрушения бетона при различных скоростях деформирования. // Бетон и железобетон. 1984. № 11.

47. Ильюшин А.А. Пластичность. – М.: Гостехиздат, 1948. – 376 с.

48. Каплун А.Б., Морозов Е.М., Олферьева М.А. ANSYS в руках инженера. Практическое руководство. – М.: Едиториал УРСС, 2003. – 272 с.

49. Карпенко Н.И. Общие модели железобетона. – М.: Стройиздат, 1996.– 416 с.

50. Карякин А.А., Оатул А.А. Расчет железобетонных балок методом конечных элементов. // Известия ВУЗов, серия « Строительство и архитектура», №3. – Новосибирск. 1977.

51. Карякин А.А. Численные методы решения задач строительства на ЭВМ: Тест лекций. – Челябинск: ЧПИ, 1989. – 47 с.

52. Карякин А.А. Расчет конструкций, зданий и сооружений с использованием персональных ЭВМ: Учебное пособие. – Челябинск: Изд-во ЮУрГУ, 2004.– 194 с.

53. Карякин А.А., Ермакова А.В. Методика учета процесса трещинообразования при расчете железобетонных конструкций методом конечных элементов. // Исследования по строительной механике и строительным конструкциям. – Челябинск: ЧПИ, 1985. – с. 131 – 133.

54. Коляскин С.Ю. Примеры расчета прямоугольных пластинок МКЭ с гарантированной погрешностью. Строительная механика и расчет сооружений. СПб.: Санкт-Петербургский государственный технический университет, 1992. – С. 88 – 96.

55. Коннор Дж., Бреббиа К. Метод конечных элементов в механике жидких сред. – Л.: Судостроение, 1979. – 264 с.

56. Крылов С. Б. Расчет железобетонных конструкций методом гладко сопряженных элементов на основе точных частных решений. – М.: Стройиздат, 2003. – 252 с.

57. Кутин Ю.Ф. Исследования сцепления с бетоном стержней периодического профиля в центрально армированных растянутых образцах. Исследования по бетону и железобетону. – Челябинск. ЧПИ, 1974. С. 133 – 141.

58. Левин В.А., Морозов Е.М., Матвиенко Ю.С. Избранные нелинейные задачи механики разрушения. – М.: ФИЗМАТЛИТ, 2004. – 408 с.

59. Максимов Ю.В., Ермакова А.В. Влияние инсоляции на работу большепролетных пространственных железобетонных конструкций // Сб. статей 4-х академических чтений. – Екатеринбург, 1998. – С. 160 – 163.

60. Максимов Ю.В., Бессонов Б.Ф., Людковский А.М. Монтаж сборной оболочки двоякой кривизны // На стройках России. – 1970. – №8. – С.26 – 28.

61. Максимов Ю.В., Бессонов Б.Ф., Карякин А.А. Расчет узла крепле-

ния мощных арматурных канатов в большепролетных железобетонных конструкциях. Исследования по строительной механике и строительным конструкциям. – Челябинск: ЧПИ, 1983. – С. 95 – 100.

62. Мандриков А.П. Примеры расчета железобетонных конструкций. – М.: Стройиздат, 1989. – 506 с.

63. Метод конечных элементов в проектировании транспортных сооружений / А. С. Городецкий, В.И. Зоворицкий, А.И. Лантух-Ляшенко, А.О. Рассказов – М: Транспорт, 1981. – 143 с.

64. Методические рекомендации по использованию библиотеки конечных элементов вычислительного комплекса «ЛИРА». – Киев: НИИАСС Госстроя УССР, 1988. – 103 с.

65. Методические рекомендации по использованию дополнительных возможностей библиотеки конечных элементов вычислительного комплекса «ЛИРА». – Киев: НИИАСС Госстроя УССР, 1988. – 151 с.

66. Миловидов В.И. Дифференцированный закон сцепления арматурных канатов К3*7(3): Исследования по бетону и железобетону. – Челябинск: ЧПИ 1969.

67. Морозов Е.М., Никишков Г.П. Метод конечных элементов в механике разрушения. – М.: Наука, 1980. – 256 с.

68. Морозов Е.М., Зернин М.В. Контактные задачи механики разрушения. – М.: Машиностроение, 1999. – 544 с.

69. Мурашкин Г.В., Снегирева А.И., Расчет железобетонных конструкций на импульсные нагрузки. Учебное пособие. – Куйбышев: КГУ, 1976. – 53 с.

70. Новое в проектировании бетонных и железобетонных конструкций / Под ред. А.А. Гвоздева. – М.: Стройиздат, 1978. – 208 с.

71. Оатул А.А., Кутин Ю.Ф., Пасечник В.В. Сцепление арматуры с бетоном (Обзор). // Известия ВУЗов, серия « Строительство и архитектура», №5. – Новосибирск. 1977.С.3–16.

72. Оатул А.А. Предложения к построению теории сцепления арматуры с бетоном // Бетон и железобетон. – № 12. – 1968. – С. 8 – 10.

73. Оатул А.А., Кутин Ю.Ф. Экспериментальное определение дифференциального закона сцепления стержневой арматуры с бетоном: Исследования по бетону и железобетону. – Челябинск: ЧПИ, 1969.

74. Оатул А.А. Расчет элементов железобетонных конструкций по двум предельным состояниям. Текст лекций. – Челябинск: ЧПИ, 1987. – Ч.2. – 64 с.

75. Оатул А.А., Максимов Ю.В., Марков В.А. и др. Опыт применения канатной арматуры на строительстве торгового центра в Челябинске // Бетон и железобетон. – 1976. – № 4. – С. 18 – 20.

76. Оден Дж. Конечные элементы в нелинейной механике сплошных сред. – М.: Мир, 1976. – 464 с.

77. Партон В.З., Морозов Е.М. Механика упругопластического разрушения. – М.: Наука, 1985. – 504 с.

78. Пестриков В.М., Морозов В.М. Механика разрушения твердых тел. Курс лекций. – СПб.: Профессия, 2002. – 320 с.

79. Пособие по проектированию бетонных и железобетонных конструкций из тяжелого бетона без предварительного напряжения арматуры (к СП 52-101-203). – М.: ООО ЦНИИПромзданий, 2005. – 214 с.

80. Пособие по проектированию бетонных и железобетонных конструкций, предназначенных для работы в условиях воздействия повышенных и высоких температур (к СНиП 2.03.01-84). – М.: ЦИТП, 1989. – 184 с.

81. Постнов В.А. Численные методы расчета судовых конструкций. – Л.: Судостроение, 1977. – 280 с.

82. Прагер В. Проблемы теории пластичности. – М.: Физматгиз, 1958. – 136 с.

83. Предельные состояния элементов железобетонных конструкций. / Под ред. С.А. Дмитриева. – М.: Стройиздат, 1976. – 216 с.

84. Программный комплекс «Мираж». Расчет конструкций на прочность. Руководство пользователя. – Киев: НИИАСС Госкомградостроительства Украины, 1994. – 434 с.

85.Программа комплекса «Лира–WINDOWS». Руководство пользователя. – Т 1–7. – Киев: НИИАСС Госкомградостроительства Украины, 1996.

86. Расчет и проектирование элементов железобетонных конструкций на основе применения ЭВМ. Конспект лекций. Оатул А.А., Карякин А.А., Кутин Ю.Ф. – Ч.4 / Под ред. Оатула А.А. – Челябинск: ЧПИ, 1980. – 67 с.

87. Ржаницын А.Р. Строительная механика.– М.: Высшая школа, 1982. – 400 с.

88. Розин Л.А. Метод конечных элементов в применении к упругим системам. – М.: Стройиздат, 1977. – 129 с.

89. Рутман Ю.Л. Метод псевдожесткостей для решения задач о предельном равновесии жестко-пластических конструкций. – СПб.: Изд-во Балтийского государственного технического университета, 1998.

90. СНиП 2.03.01-84. Бетонные и железобетонные конструкции. – М.: ЦИТП Госстроя СССР, 1985. – 79 с.

91. СНиП 52-01-2003. Бетонные и железобетонные конструкции. – М.: ФГУП ЦПП, 2004. – 26 с.

92. СНиП 2.01.07-85. Нагрузки и воздействия. – М.: ЦИТП Госстроя СССР, 1985.

93. Снитко Н.К. Строительная механика. – М.: Высшая школа, 1980. – 431 с.

94 Сонин С.А. Некоторые результаты расчета сборно-монолитной балки МКЭ. // Исследования по строительной механике и строительным конструкциям. Тематический сборник научных трудов. – Челябинск. 1987.С.82–85.

95. Сонин С.А., Карякин А.А. Экспериментальные и теоретические исследования сборно-монолитных балок таврового сечения. Исследования по строительной механике грунтов. – Челябинск: ЧПИ, 1979. – С. 144–154.

96. Улицкий И.И., Ривкин С.А., Самолетов М.В. и др. Железобетонные конструкции. – Киев: Будивельник, 1973. – 992 с.

97. Усиленные наружные стеновые панели. Усиленные внутренние стеновые панели. (Разработаны СУ-5 ПСМО «Челябинскгражданстрой» для 10 этажных блок-секций. Вариант свайных фундаментов с нижним

безростверковым основанием). Альбом Ч.10 р.2-1а. Серия 121.– Челябинск: 1986. – 27 с.

98. Филин А.П. Элементы теории оболочек. – Л.: Стройиздат, 1975. – 256 с.

99. Хайдуков Г.К., Качановский Е.К., Пятикрестовский К.П. Испытания сборной оболочки покрытия размерами в плане 102х102 м // Бетон и железобетон. – 1976. – № 4. – С. 12 – 14.

100. Хилл Р. Математическая теория пластичности. – М.: Гостехиздат, 1956. –407 с.

101. Ходж Ф.Г. Расчет конструкций с учетом пластических деформаций. – М.: Машгиз, 1963. – 380 с.

102. Холмянский М.М. Закладные детали сборных элементов. – М.: Стройиздат, 1968.

103. Черкасов В.А., Важенин В.А. Монтаж сборной железобетонной оболочки размерами 102×102м // Реферативная информация о передовом опыте. Серия VII. Изготовление металлических и монтаж строительных конструкций. – М.: ЦБНТИ, 1975. – Вып.7(76). – С. 6 – 8.

104. Черный А.С., Пушкарев Л., Шапиро А.В. Монтаж сборной оболочки двоякой кривизны // На стройках России. – 1970. – № 8. – С. 26–28.

105. Черный А.С., Максимов Ю.В., Молодцов М.В. Возведение большепролетной преднапряженной сборно-монолитной оболочки торгового центра в г. Челябинске: Учебное пособие / Под ред. Ю.В. Максимова. – Челябинск: ЮУрГУ, 2004. – 44 с.

106. Шапиро А.В., Лобанов Н.Д., Черный А.С. Сборная железобетонная оболочка положительной кривизны размером 102×102 м в Челябинске // Бетон и железобетон. – 1973. – № 7. – С. 9 – 11.

107. Шапиро А.В., Пятикрестовский К.П., Качановский Е.К. Железобетонная оболочка размерами 102×102 м с предварительно напряженным контуром. Реферативный сборник. Общие вопросы строительства. Отечественный опыт. – М.: ЦИНИС, 1973. – Вып.1. – С. 25–29.

108. Шугаев В.В. Инженерные методы в нелинейной теории предельного равновесия оболочек. – М.: Готика, 2001. – 368 с.

109. Clark D.A. Mathematics for Engineering: An Active Learning Approach – London: DP, 1994. – 294 p.

110. Ermakova A. Allowance for non-linear properties of reinforced concrete structures in finite element analysis. Concrete for Extreme Conditions. Proceedings of the International Conference held at the University of Dundee, Scotland, UK on 911 September 2002. – London: Thomas Telford Publishing, 2002. – p. 397–404.

111. Ermakova A, Karyakin A. Nonlinear Finite Element Analysis of Load-Bearing Wall Panels. Proceedings of the Forth International Conference on Concrete under Severe Conditions, CONSEC'04, June 2023 2004. – Seoul, 8 p.

112. Ermakova A. Limit State Analysis of Reinforced Concrete Structures by Finite Element Method. Application of Codes, Design and Regulations. Proceedings of the International Conference held at the University of Dundee, Scotland,

UK on 5-7 July 2005. – London: Thomas Telford Publishing, 2005. p. 587–594.

113. Heyman J. Beams and framed structures. – Oxford & oth.: Pergamon Press, 1974. –136 p.

114. Kong F.K., Evans R.H. Reinforced and Prestressed Concrete. 2-nd ed. Thomas Nelson and Sons ltd. Pitman Press, GB, 1980. – 412 p.

115. Livesley R.K. Matrix Methods of Structural Analysis. – Oxford & oth.: Pergamon Press, 1975, XII. – 227 p.

116. Marshall W.T., Nelson H.M. Structures. SI units – 2-ed. – London: Pitmen, 1978. – 455 p.

117. Maximov Yu. V., Ermakova A.V. Analyses of Chelyabinsk shopping centre shell under thermal, static and dynamic loads. Spatial structures in new and renovation project of buildings and construction. International congress ICSS-98. June 22–26 1998. – Moscow: 'Construction" State Research Centre of Russia. 1998, volume 1. p. 368–373.

118. Maximov I., Ermakova A. Necessity of Allowance for Daily Temperature Difference in Design and Operation of Long-Span Shells. Proceedings of the Third International Conference on Concrete under Severe Conditions, CONSEC'01, June 18–20 2001. – Vancouver, BC, Canada, volume 1, p. 1017–1023.

119. Nayak G.C., Zienkiewicz O.C. Elastic-plastic stress analysis. A generalization for various constitutive relations including strain shortening. – Int. J. Numer. Meth. Eng., 1973, V. 5, № 1. – P. 113–135.

120. Ngo D. and Scordelis A.C. Finite Element Analysis of Reinforced Concrete Beams // ACI Journal. – 1967. – V. 64, № 3.

121. Nilson A.H. Nonlinear Analysis of Reinforced Concrete by Finite Element Method // ACI Journal. – 1967. – V. 65, № 9. – P. 757 – 66.

122. Scordelis A.C. Computer Models in Nonlinear of Reinforced and Prestressed Concrete Structures // J. of Prestressed Concrete Inst. – 1984. – V. 29, №6. – P. 116 –132.

123. Stiffness and deflection analysis of complex structures. Turner M.J., Clough R.W., Martin H.C., Topp L.J // J. Aeronaut Science. – 1956. – V. 23, №9. – P. 805 – 824.

Anna Vitalievna **Ermakova**

ADDITIONAL FINITE ELEMENT METHOD FOR ANALYSIS OF REINFORCED CONCRETE STRUCTURES AT LIMIT STATES

Translated from the Russian by *O.V. Ermakova*
Editor *O.A. Taranova*
Computer-aided page proof by *Ya.P. Yasina*
Cover design by *N.S.Romanova*

Signed for printing 12.01.2016. Format 60×90 1/16.
Offset paper. Times type. Offset printing.
Conventional 7.25 printed sheets.

ASV Constraction, Sweden,
Mårdvägen 16 131 50 Saltsjö-Duvnäs

www.ingramcontent.com/pod-product-compliance
Lightning Source LLC
Chambersburg PA
CBHW052016230326
41598CB00078B/3492
9789198222326